THE
PHYSICS OF
SOLAR CELLS

THE
PHYSICS OF
SOLAR CELLS

Jenny Nelson
Imperial College, UK

Imperial College Press

ICP

Published by

Imperial College Press
57 Shelton Street
Covent Garden
London WC2H 9HE

Distributed by

World Scientific Publishing Co. Pte. Ltd.
5 Toh Tuck Link, Singapore 596224
USA office: Suite 202, 1060 Main Street, River Edge, NJ 07661
UK office: 57 Shelton Street, Covent Garden, London WC2H 9HE

British Library Cataloguing-in-Publication Data
A catalogue record for this book is available from the British Library.

ISBN   1-86094-340-3
ISBN   1-86094-349-7 (pbk)

Printed in Singapore by Mainland Press

Preface

Over the last ten years, photovoltaics has emerged to become an application of recognised potential and has attracted the interest of increasing numbers of students and researchers. The purpose of this book is to provide an introduction to, and overview of, the physics of the photovoltaic cell. It should be suitable for undergraduate physicists and engineers who are interested in this application of semiconductor physics, and to non-specialist graduates and others who require a background in the physical principles of solar cells. The focus is on the basic semiconductor physics relevant to photovoltaics, physical models of photovoltaic devices and how these relate to the design and function of practical devices. It should enable the reader to understand how solar cells work, to understand the concepts and models of solar cell device physics, and to formulate and solve relevant physical problems. Although practical materials and device designs are used as examples, the book is not intended as a comprehensive review of photovoltaic materials and devices, nor of the latest developments in photovoltaics research.

Chapter 1 introduces the solar cell as a simple current generator and defines the performance characteristics which are used to describe and compare solar cells. Chapter 2 describes in general terms how light energy is converted into electricity, comparing the photo*voltaic* converter with other systems and evaluating the limits to efficiency. Chapters 3 and 4 cover the basic physics of the semiconductor, the semiconductor transport equations and the processes of light absorption and carrier recombination. Chapter 5 focuses on the concept of the asymmetric junction, and details the different types of junction which are exploited in photovoltaics. Chapter 6 applies the theory of earlier chapters to a p–n junction, the classical model of a solar cell. Chapters 7 and 8 are concerned with the range of photovoltaic materials and device designs. Chapter 7 deals with monocrystalline p–n junction devices, relating the model of Chapter 6 to practical devices, using crystalline silicon and gallium arsenide cells as examples. Chapter 8 deals with thin film photovoltaic materials, discussing physical processes and design issues relevant to thin films and focusing on the ways in which the standard model must be adapted for thin film devices. Chapter 9 deals with various techniques for managing light in order to maximise performance, and Chapter 10 covers a range of approaches, mainly theoretical, to increasing the efficiency of solar cells above the limit for a single band gap photoconverter.

I am grateful to all of the people who have helped me prepare this book. In particular, to Keith Barnham for passing the original proposal from Imperial College Press in my direction; to Leon Freris and David Infield for giving me the opportunity to teach the physics of solar cells to MSc students at Loughborough, and so establish the basic course from which this book developed; to all the research students in photovoltaics at Imperial College for raising so many interesting questions, especially Jenny Barnes, James Connolly and Benjamin Kluftinger; to Ralph Gottshalg, Tom Markvart and Peter Wuerfel for help with questions related to material in this book; to Ned Ekins-Daukes and Jane Nelson for their helpful comments on the text; to Clare Nelson for the cover illustration and to all other colleagues who have helped in my endeavours to understand how these things work, in particular to Richard Corkish, James Durrant, Michael Grätzel, Martin Green, Christiana Honsberg, Stefan Kettemann and Ellen Moons. I am grateful to the Greenpeace Environmental Trust for funding me to study solar cells before they were popular, and to the UK Engineering and Physical Sciences Research Council and for an Advanced Research Fellowship which allowed me to spend my Saturday afternoons writing chapters instead of lectures. Finally I am grateful to John Navas for his encouragement to start on this project and to Laurent Chaminade and his staff at IC Press and to Lakshmi Narayan and colleagues at World Scientific, for their help in seeing it through.

This book is dedicated to the memory of Stephen Robinson and M.V. McCaughan.

Jenny Nelson
London, April 2002

Contents

Fundamental constants

h Planck's constant
$\hbar$ Planck's constant$/2\pi$
ε_0 dielectric permittivity of free space
σ_B Stefan's constant
c speed of light in vacuum
k_B Boltzmann's constant
m_0 free electron mass
q charge on the electron

Symbols used in the text

α absorption coefficient
β spectral photon flux density per unit solid angle
χ electron affinity
$\Delta\mu$ quasi Fermi level separation or chemical potential of light
ε emissivity *i.e.* probability of photon emission
ε_s dielectric permittivity of semiconductor
ε polarisation vector of light
ϕ electrostatic potential
Φ work function
ϕ_0 neutrality level
η power conversion efficiency
κ_s imaginary part of refractive index of semiconductor
λ wavelength of light
μ chemical potential
$\mu_n; \mu_p$ electron mobility; hole mobility
ν frequency of light
θ_c critical angle at optical interface
θ_{sun} angular width of the sun
ρ charge density or resistivity
σ conductivity
τ lifetime
$\tau_n; \tau_p$ electron lifetime; hole lifetime
ω angular frequency of light
Ω solid angle
ψ wavefunction
∇, ∇_r grad operator with respect to position
∇_k grad operator with respect to wavevector

a	probability of photon absorption
A	cell area
b	spectral photon flux density normal to surface
$B_n; B_p$	coefficient for bimolecular capture by trap of electrons; holes
B_{rad}	coefficient of bimolecular radiative recombination
D	diffusion coefficient
$D_n; D_p$	diffusion coefficient of electron; diffusion coefficient of hole
E	energy
E_0	electromagnetic field strength
E_F	Fermi energy or Fermi level
$E_{F_n}; E_{F_p}$	electron quasi-Fermi energy level; hole quasi-Fermi level
E_g	band gap
E_t	energy of trap state
E_v	energy of valence band edge
E_c	energy of conduction band edge
E_i	intrinsic energy level
E_{vac}	vacuum energy level
F	electrostatic field
F	force
$f(\mathbf{k}, \mathbf{r})$	probability occupation function for electronic state at $\mathbf{k}, \mathbf{r}$
f_0	Fermi Dirac probability occupation function
$F_a; F_e; F_s; F_X$	geometrical factor relating normal to angular photon flux density for emission from: ambient; cell; sun; concentrated light source
FF	fill factor
g	spectral photogeneration rate per unit volume
G	generation rate per unit volume
G_n	electron generation rate per unit volume
G_p	hole generation rate per unit volume
$g(E)$	density of electronic states per unit energy per unit crystal volume
H	Hamiltonian operator
I	current
I_0	incident light intensity
J	current density
J_{sc}	short circuit current density
J_{dark}	dark current density
$J_n; J_p$	electron current density; hole current density

j_n; j_p	spectral electron current density; spectral hole current density
k	crystal wavevector
L	diffusion length or length of crystal sample
L_n; L_p	diffusion length of electron; diffusion length of hole
m	diode ideality factor
m_c^*	conduction band effective mass
m_v^*	valence band effective mass
M	dipole matrix element
n	density of electrons per unit volume
n_0	equilibrium electron density
n_i	intrinsic carrier density
n_s	refractive index of semiconductor
N_a	density of acceptor impurity atoms
N_d	density of donor impurity atoms
N_c	effective conduction band density of states
N_v	effective valence band density of states
N_i	charged background doping in intrinsic layer
N_t	density of trap states
N_s	density of interface states per unit area
p	density of holes per unit volume
p	momemtum
P	power density
p_0	equilibrium hole density
Q	charge
QE	quantum efficiency
r	position
r	transition rate
R	reflectivity
R_s	series resistance
R_{sh}	shunt or parallel resistance
s	vector defining a point on surface
S_n	electron surface recombination velocity
S_p	hole surface recombination velocity
t	time
T	temperature
T_s	temperature of sun
U_E	energy density of radiation per unit volume
U	recombination rate per unit volume
U_n	electron recombination rate per unit volume

U_{p}	hole recombination rate per unit volume
U_{Aug}	Auger recombination rate
U_{rad}	radiative recombination rate
U_{SRH}	Shockley Read Hall recombination rate
$\mathcal{V}$	volume
v	velocity
V	voltage or bias
V_{bi}	built in bias
V_{oc}	open circuit voltage
V_{n}	donor ionisation energy
V_{p}	acceptor ionisation energy
$w_{\mathrm{n}}; w_{\mathrm{p}}$	thickness of depletion region in n layer; in p layer
X	concentration factor

Acronyms

e.m.f.	electromotive force
ac	alternating current
dc	direct current
AM	air mass
AM1.5	air mass 1.5 spectrum — standard for solar cell calibration
STC	standard test conditions
CB	conduction band
VB	valence band
FGR	Fermi's golden rule
PV	photovoltaics
a-Si	amorphous silicon
c-Si	crystalline silicon
μ-Si	microcrystalline silicon
a-Si:H	hydrogenated amorphous silicon
a-SiC	amorphous silicon-carbon alloy
a-SiGe	amorphous silicon-germanium alloy
DOS	density of states
JDOS	joint density of states
i region	intrinsic or undoped region of a p–i–n junction
SCR	space charge region
QE	quantum efficiency
SRH	Shockley–Read–Hall
GaAs	gallium arsenide
CdTe	cadmium telluride

CIGS   copper indium gallium diselenide
AR   antireflection
PERL   passivated emitter, rear locally diffused solar cell
PESC   passivated emitter solar cell
TCO   transparent conducting oxide
PR   photon recycling
QD   quantum dot
QW   quantum well
LO   longitudinal optical (of phonons)

Chapter 1

Introduction

1.1. Photons In, Electrons Out: The Photovoltaic Effect

Solar photovoltaic energy conversion is a one-step conversion process which
generates electrical energy from light energy. The explanation relies on ideas
from quantum theory. Light is made up of packets of energy, called *photons*,
whose energy depends only upon the frequency, or colour, of the light. The
energy of visible photons is sufficient to excite electrons, bound into solids,
up to higher energy levels where they are more free to move. An extreme
example of this is the photoelectric effect, the celebrated experiment which
was explained by Einstein in 1905, where blue or ultraviolet light provides
enough energy for electrons to escape completely from the surface of a
metal. Normally, when light is absorbed by matter, photons are given up to
excite electrons to higher energy states within the material, but the excited
electrons quickly relax back to their ground state. In a photovoltaic device,

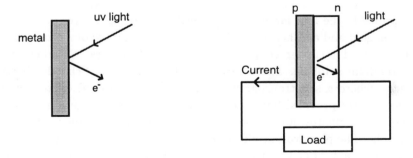

Fig. 1.1.   Comparison of the photoelectric effect (left), where uv light liberates electrons
from the surface of a metal, with the photovoltaic effect in a solar cell (right). The
photovoltaic cell needs to have some spatial asymmetry, such as contacts with different
electronic properties, to drive the excited electrons through the external circuit.

however, there is some built-in asymmetry which pulls the excited electrons away before they can relax, and feeds them to an external circuit. The extra energy of the excited electrons generates a *potential difference*, or electromotive force (e.m.f.). This force drives the electrons through a load in the external circuit to do electrical work.

The effectiveness of a photovoltaic device depends upon the choice of light absorbing materials and the way in which they are connected to the external circuit. The following chapters will deal with the underlying physical ideas, the device physics of solar cells, the properties of photovoltaic materials and solar cell design. In this chapter we will summarise the main characteristics of a photovoltaic cell without discussing its physical function in detail.

1.2.   Brief History of the Solar Cell

The photovoltaic effect was first reported by Edmund Bequerel in 1839 when he observed that the action of light on a silver coated platinum electrode immersed in electrolyte produced an electric current. Forty years later the first solid state photovoltaic devices were constructed by workers investigating the recently discovered photoconductivity of selenium. In 1876 William Adams and Richard Day found that a photocurrent could be produced in a sample of selenium when contacted by two heated platinum contacts. The photovoltaic action of the selenium differed from its photoconductive action in that a current was produced spontaneously by the action of light. No external power supply was needed. In this early photovoltaic device, a rectifying junction had been formed between the semiconductor and the metal contact. In 1894, Charles Fritts prepared what was probably the first large area solar cell by pressing a layer of selenium between gold and another metal. In the following years photovoltaic effects were observed in copper–copper oxide thin film structures, in lead sulphide and thallium sulphide. These early cells were thin film Schottky barrier devices, where a semitransparent layer of metal deposited on top of the semiconductor provided both the asymmetric electronic junction, which is necessary for photovoltaic action, and access to the junction for the incident light. The photovoltaic effect of structures like this was related to the existence of a barrier to current flow at one of the semiconductor–metal interfaces (*i.e.*, rectifying action) by Goldman and Brodsky in 1914. Later, during the 1930s, the theory of metal–semiconductor barrier layers was developed by Walter Schottky, Neville Mott and others.

However, it was not the photovoltaic properties of materials like selenium which excited researchers, but the photoconductivity. The fact that the current produced was proportional to the intensity of the incident light, and related to the wavelength in a definite way meant that photoconductive materials were ideal for photographic light meters. The photovoltaic effect in barrier structures was an added benefit, meaning that the light meter could operate without a power supply. It was not until the 1950s, with the development of good quality silicon wafers for applications in the new solid state electronics, that potentially useful quantities of power were produced by photovoltaic devices in crystalline silicon.

In the 1950s, the development of silicon electronics followed the discovery of a way to manufacture p–n junctions in silicon. Naturally n type silicon wafers developed a p type skin when exposed to the gas boron trichloride. Part of the skin could be etched away to give access to the n type layer beneath. These p–n junction structures produced much better rectifying action than Schottky barriers, and better photovoltaic behaviour. The first silicon solar cell was reported by Chapin, Fuller and Pearson in 1954 and converted sunlight with an efficiency of 6%, six times higher than the best previous attempt. That figure was to rise significantly over the following years and decades but, at an estimated production cost of some $200 per Watt, these cells were not seriously considered for power generation for several decades. Nevertheless, the early silicon solar cell did introduce the possibility of power generation in remote locations where fuel could not easily be delivered. The obvious application was to satellites where the requirement of reliability and low weight made the cost of the cells unimportant and during the 1950s and 60s, silicon solar cells were widely developed for applications in space.

Also in 1954, a cadmium sulphide p–n junction was produced with an efficiency of 6%, and in the following years studies of p–n junction photovoltaic devices in gallium arsenide, indium phosphide and cadmium telluride were stimulated by theoretical work indicating that these materials would offer a higher efficiency. However, silicon remained and remains the foremost photovoltaic material, benefiting from the advances of silicon technology for the microelectronics industry. Short histories of the solar cell are given elsewhere [Shive, 1959; Wolf, 1972; Green, 1990].

In the 1970s the crisis in energy supply experienced by the oil-dependent western world led to a sudden growth of interest in alternative sources of energy, and funding for research and development in those areas. Photovoltaics was a subject of intense interest during this period, and a range of strategies for producing photovoltaic devices and materials more cheaply and for improving device efficiency were explored. Routes to lower cost included photoelectrochemical junctions, and alternative materials such as polycrystalline silicon, amorphous silicon, other 'thin film' materials and organic conductors. Strategies for higher efficiency included tandem and other multiple band gap designs. Although none of these led to widespread commercial development, our understanding of the science of photovoltaics is mainly rooted in this period.

During the 1990s, interest in photovoltaics expanded, along with growing awareness of the need to secure sources of electricity alternative to fossil fuels. The trend coincides with the widespread deregulation of the electricity markets and growing recognition of the viability of decentralised power. During this period, the economics of photovoltaics improved primarily through economies of scale. In the late 1990s the photovoltaic production expanded at a rate of 15–25% per annum, driving a reduction in cost. Photovoltaics first became competitive in contexts where conventional electricity supply is most expensive, for instance, for remote low power applications such as navigation, telecommunications, and rural electrification and for enhancement of supply in grid-connected loads at peak use [Anderson, 2001]. As prices fall, new markets are opened up. An important example is building integrated photovoltaic applications, where the cost of the photovoltaic system is offset by the savings in building materials.

1.3.   Photovoltaic Cells and Power Generation

1.3.1.   *Photovoltaic cells, modules and systems*

The solar *cell* is the basic building block of solar photovoltaics. The cell can be considered as a two terminal device which conducts like a diode in the dark and generates a photovoltage when charged by the sun. Usually it is a thin slice of semiconductor material of around 100 cm^2 in area. The surface is treated to reflect as little visible light as possible and appears dark blue or black. A pattern of metal contacts is imprinted on the surface to make electrical contact (Fig. 1.2(a)).

When charged by the sun, this basic unit generates a dc photovoltage of 0.5 to 1 volt and, in short circuit, a photocurrent of some tens of milliamps per cm^2. Although the current is reasonable, the voltage is too small for most applications. To produce useful dc voltages, the cells are connected together in series and encapsulated into *modules*. A module typically contains

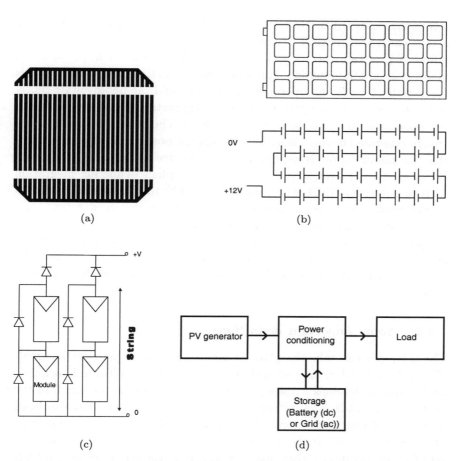

Fig. 1.2. (a) Photovoltaic cell showing surface contact patterns (b) In a module, cells are usually connected in series to give a standard dc voltage of 12 V (c) For any application, modules are connected in series into strings and then in parallel into an array, which produces sufficient current and voltage to meet the demand. (d) In most cases the photovoltaic array should be integrated with components for charge regulation and storage.

28 to 36 cells in series, to generate a dc output voltage of 12 V in standard illumination conditions (Fig. 1.2(b)). The 12 V modules can be used singly, or connected in parallel and series into an array with a larger current and voltage output, according to the power demanded by the application (Fig. 1.2(c)). Cells within a module are integrated with bypass and blocking diodes in order to avoid the complete loss of power which would result if one

cell in the series failed. Modules within arrays are similarly protected. The array, which is also called a photovoltaic *generator*, is designed to generate power at a certain current and a voltage which is some multiple of 12 V, under standard illumination. For almost all applications, the illumination is too variable for efficient operation all the time and the photovoltaic generator must be integrated with a charge storage system (a battery) and with components for power regulation (Fig. 1.2(d)). The battery is used to store charge generated during sunny periods and the power conditioning ensures that the power supply is regular and less sensitive to the solar irradiation. For ac electrical power, to power ac designed appliances and for integration with an electricity grid, the dc current supplied by the photovoltaic modules is converted to ac power of appropriate frequency using an *inverter*.

The design and engineering of photovoltaic systems is beyond the scope of this book. A more detailed introduction is given by Markvart [Markvart, 2000] and Lorenzo [Lorenzo, 1994]. Photovoltaic systems engineering depends to a large degree upon the electrical characteristics of the individual cells.

1.3.2. *Some important definitions*

The solar cell can take the place of a battery in a simple electric circuit (Fig. 1.3). In the dark the cell in circuit A does nothing. When it is switched on by light it develops a voltage, or e.m.f., analogous to the e.m.f. of the battery in circuit B. The voltage developed when the terminals are isolated (infinite load resistance) is called the *open circuit voltage* V_{oc}. The current drawn when the terminals are connected together is the *short circuit current* I_{sc}. For any intermediate load resistance R_L the cell develops a voltage V between 0 and V_{oc} and delivers a current I such that $V = IR_L$ and $I(V)$ is determined by the *current–voltage characteristic* of the cell under

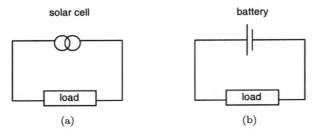

Fig. 1.3.   The solar cell may replace a battery in a simple circuit.

that illumination. Thus both I and V are determined by the illumination as well as the load. Since the current is roughly proportional to the illuminated area, the *short circuit current density* J_{sc} is the useful quantity for comparison. These quantities are defined for a simple, ideal diode model of a solar cell in Sec. 1.4 below.

Box 1.1. Solar cell compared with conventional battery

The photovoltaic cell differs from a simple dc battery in these respects: the e.m.f. of the battery is due to the permanent electrochemical potential difference between two phases in the cell, while the solar cell derives its e.m.f. from a temporary change in electrochemical potential caused by light. The power delivered by the battery to a constant load resistance is relatively constant, while the power delivered by the solar cell depends on the incident light intensity, and not primarily on the load (Fig. 1.4). The battery is completely discharged when it reaches the end of its life, while the solar cell, although its output varies with intensity, is in principle never exhausted, since it can be continually recharged with light.

The battery is modelled electrically as a *voltage generator* and is characterised by its e.m.f. (which, in practice, depends upon the degree of discharge), its charge capacity, and by a polarisation curve which describes how the e.m.f. varies with current [Vincent 1997]. The solar cell, in contrast, is better modelled as a *current generator*, since for all but the largest loads the current drawn is independent of load. But its characteristics depend entirely on the nature of the illuminating source, and so I_{sc} and V_{oc} must be quoted for a known spectrum, usually for standard test conditions (defined below).

1.4. Characteristics of the Photovoltaic Cell: A Summary

1.4.1. *Photocurrent and quantum efficiency*

The photocurrent generated by a solar cell under illumination at short circuit is dependent on the incident light. To relate the photocurrent density, J_{sc}, to the incident spectrum we need the cell's *quantum efficiency*, (QE). $QE(E)$ is the probability that an incident photon of energy E will deliver one electron to the external circuit. Then

$$J_{sc} = q \int b_s(E) QE(E) dE \tag{1.1}$$

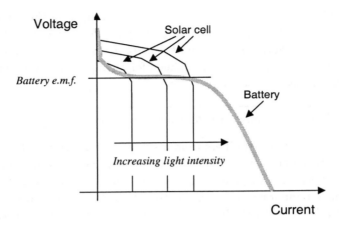

Fig. 1.4.   Voltage–current curves of a conventional battery (grey) and a solar cell under different levels of illumination. A battery normally delivers a constant e.m.f. at different levels of current drain except for very low resistance loads, when the e.m.f. begins to fall. The battery e.m.f. will also deteriorate when the battery is heavily discharged. The solar cell delivers a constant current for any given illumination level while the voltage is determined largely by the resistance of the load. For photovoltaic cells it is usual to plot the data in the opposite sense, with current on the vertical axis and voltage on the horizontal axis. This is because the photovoltaic cell is essentially a current source, while the battery is a voltage source.

where $b_s(E)$ is the incident spectral photon flux density, the number of photons of energy in the range E to $E + dE$ which are incident on unit area in unit time and q is the electronic charge. QE depends upon the absorption coefficient of the solar cell material, the efficiency of charge separation and the efficiency of charge collection in the device but does not depend on the incident spectrum. It is therefore a key quantity in describing solar cell performance under different conditions. Figure 1.5 shows a typical QE spectrum in comparison with the spectrum of solar photons.

QE and spectrum can be given as functions of either photon energy or wavelength, λ. Energy is a more convenient parameter for the physics of solar cells and it will be used in this book. The relationship between E and λ is defined by

$$E = \frac{hc}{\lambda} \qquad (1.2)$$

where h is Planck's constant and c the speed of light in vacuum. A convenient rule for converting between photon energies, in electron-Volts, and wavelengths, in nm, is $E/\text{eV} = 1240/(\lambda/\text{nm})$.

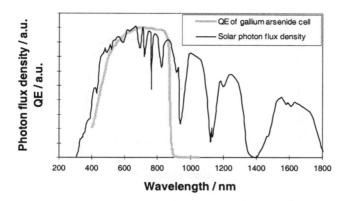

Fig. 1.5. Quantum efficiency of GaAs cell compared to the solar spectrum. The vertical scale is in arbitrary units, for comparison. The short circuit photocurrent is obtained by integrating the product of the photon flux density and QE over photon energy. It is desirable to have a high QE at wavelengths where the solar flux density is high.

1.4.2. *Dark current and open circuit voltage*

When a load is present, a potential difference develops between the terminals of the cell. This potential difference generates a current which acts in the opposite direction to the photocurrent, and the net current is reduced from its short circuit value. This reverse current is usually called the *dark current* in analogy with the current $I_{\mathrm{dark}}(V)$ which flows across the device under an applied voltage, or *bias*, V in the dark. Most solar cells behave like a diode in the dark, admitting a much larger current under forward bias ($V > 0$) than under reverse bias ($V < 0$). This rectifying behaviour is a feature of photovoltaic devices, since an asymmetric junction is needed to achieve charge separation. For an ideal diode the *dark current density* $J_{\mathrm{dark}}(V)$ varies like

$$J_{\mathrm{dark}}(V) = J_{\mathrm{o}}(e^{qV/k_{\mathrm{B}}T} - 1) \qquad (1.3)$$

where J_{o} is a constant, k_{B} is Boltzmann's constant and T is temperature in degrees Kelvin.

The overall current voltage response of the cell, its *current–voltage characteristic*, can be approximated as the sum of the short circuit photocurrent and the dark current (Fig. 1.6). This step is known as the *superposition* approximation. Although the reverse current which flows in reponse to voltage in an illuminated cell is not formally equal to the current which flows in the dark, the approximation is reasonable for many photovoltaic materials and

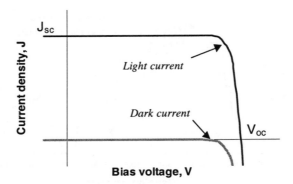

Fig. 1.6. Current–voltage characteristic of ideal diode in the light and the dark. To a first approximation, the net current is obtained by shifting the bias dependent dark current up by a constant amount, equal to the short circuit photocurrent. The sign convention is such that the short circuit photocurrent is positive.

will be used for the present discussion. The sign convention for current and voltage in photovoltaics is such that the photocurrent is positive. This is the opposite to the usual convention for electronic devices. With this sign convention the net current density in the cell is

$$J(V) = J_{\text{sc}} - J_{\text{dark}}(V)\,, \tag{1.4}$$

which becomes, for an ideal diode,

$$J = J_{\text{sc}} - J_0(e^{q V/k_{\text{B}} T} - 1)\,. \tag{1.5}$$

This result is derived in Chapter 6.

When the contacts are isolated, the potential difference has its maximum value, the open circuit voltage V_{oc}. This is equivalent to the condition when the dark current and short circuit photocurrent exactly cancel out. For the ideal diode, from Eq. 1.5,

$$V_{\text{oc}} = \frac{kT}{q} \ln\left(\frac{J_{\text{sc}}}{J_0} + 1\right)\,. \tag{1.6}$$

Equation 1.6 shows that V_{oc} increases logarithmically with light intensity. Note that voltage is defined so that the photovoltage occurs in forward bias, where $V > 0$.

Figure 1.6 shows that the current–voltage product is positive, and the cell generates power, when the voltage is between 0 and V_{oc}. At $V < 0$, the illuminated device acts as a photodetector, consuming power to generate a

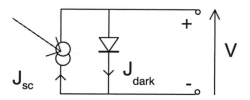

Fig. 1.7.   Equivalent circuit of ideal solar cell.

photocurrent which is light dependent but bias independent. At $V > V_{oc}$, the device again consumes power. This is the regime where light emitting diodes operate. We will see later that in some materials the dark current is accompanied by the emission of light.

Electrically, the solar cell is equivalent to a current generator in parallel with an asymmetric, non linear resistive element, *i.e.*, a diode (Fig. 1.7). When illuminated, the ideal cell produces a photocurrent proportional to the light intensity. That photocurrent is divided between the variable resistance of the diode and the load, in a ratio which depends on the resistance of the load and the level of illumination. For higher resistances, more of the photocurrent flows through the diode, resulting in a higher potential difference between the cell terminals but a smaller current though the load. The diode thus provides the photovoltage. Without the diode, there is nothing to drive the photocurrent through the load.

1.4.3.   *Efficiency*

The operating regime of the solar cell is the range of bias, from 0 to V_{oc}, in which the cell delivers power. The cell *power density* is given by

$$P = JV .\tag{1.7}$$

P reaches a maximum at the cell's operating point or *maximum power point*. This occurs at some voltage V_m with a corresponding current density J_m, shown in Fig. 1.8. The optimum load thus has sheet resistance given by V_m/J_m. The *fill factor* is defined as the ratio

$$FF = \frac{J_m V_m}{J_{sc} V_{oc}}\tag{1.8}$$

and describes the 'squareness' of the J–V curve.

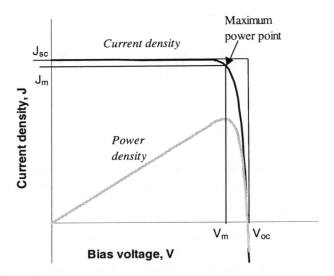

Fig. 1.8. The current voltage (black) and power–voltage (grey) characteristics of an ideal cell. Power density reaches a maximum at a bias V_m, close to V_{oc}. The maximum power density $J_m \times V_m$ is given by the area of the inner rectangle. The outer rectangle has area $J_{sc} \times V_{oc}$. If the fill factor were equal to 1, the current voltage curve would follow the outer rectangle.

The *efficiency* η of the cell is the power density delivered at operating point as a fraction of the incident light power density, P_s,

$$\eta = \frac{J_m V_m}{P_s}. \tag{1.9}$$

Efficiency is related to J_{sc} and V_{oc} using FF,

$$\eta = \frac{J_{sc} V_{oc} FF}{P_s}. \tag{1.10}$$

These four quantities: J_{sc}, V_{oc}, FF and η are the key performance characteristics of a solar cell. All of these should be defined for particular illumination conditions. The Standard Test Condition (STC) for solar cells is the Air Mass 1.5 spectrum, an incident power density of 1000 W m^{-2}, and a temperature of 25°C. (Standard and other solar spectra are discussed in Chapter 2.) The performance characteristics for the most common solar cell materials are listed in Table 1.1.

Table 1.1 shows that solar cell materials with higher J_{sc} tend to have lower V_{oc}. This is a consequence of the material used, and particularly of

Table 1.1. Performance of some types of PV cell [Green *et al.*, 2001].

Cell Type	Area (cm^2)	V_{oc} (V)	J_{sc} (mA/cm^2)	FF	Efficiency (%)
crystalline Si	4.0	0.706	42.2	82.8	24.7
crystalline GaAs	3.9	1.022	28.2	87.1	25.1
poly-Si	1.1	0.654	38.1	79.5	19.8
a-Si	1.0	0.887	19.4	74.1	12.7
CuInGaSe$_2$	1.0	0.669	35.7	77.0	18.4
CdTe	1.1	0.848	25.9	74.5	16.4

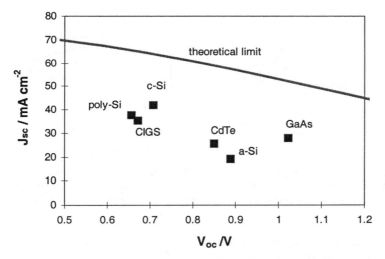

Fig. 1.9. Plot of J_{sc} against V_{oc} for the cells listed in Table 1.1. Materials with high V_{oc} tend to have lower J_{sc}. This is due to the band gap of the semiconductor material. The grey line shows the relationship expected in the theoretical limit.

the band gap of the semiconductor. In Chapter 2 we will see that there is a fundamental compromise between photocurrent and voltage in photovoltaic energy conversion. Figure 1.9 illustrates the correlation between J_{sc} and V_{oc} for the cells in Table 1.1, together with the relationship for a cell of maximum efficiency.

1.4.4. *Parasitic resistances*

In real cells power is dissipated through the resistance of the contacts and through leakage currents around the sides of the device. These effects are

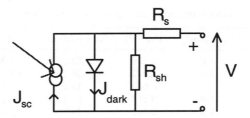

Fig. 1.10.   Equivalent circuit including series and shunt resistances.

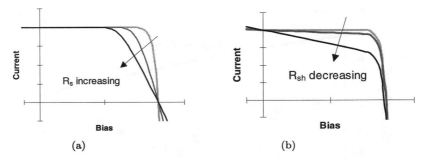

Fig. 1.11.   Effect of (a) increasing series and (b) reducing parallel resistances. In each case the outer curve has $R_s = 0$ and $R_{sh} = \infty$. In each case the effect of the resistances is to reduce the area of the maximum power rectangle compared to $J_{sc} \times V_{oc}$.

equivalent electrically to two parasitic resistances in series (R_s) and in parallel (R_{sh}) with the cell (Fig. 1.10).

The series resistance arises from the resistance of the cell material to current flow, particularly through the front surface to the contacts, and from resistive contacts. Series resistance is a particular problem at high current densities, for instance under concentrated light. The parallel or shunt resistance arises from leakage of current through the cell, around the edges of the device and between contacts of different polarity. It is a problem in poorly rectifying devices.

Series and parallel resistances reduce the fill factor as shown in Fig. 1.11. For an efficient cell we want R_s to be as small and R_{sh} to be as large as possible.

When parasitic resistances are included, the diode equation becomes

$$J = J_{sc} - J_o(e^{q(V+JAR_s)/kT} - 1) - \frac{V + JAR_s}{R_{sh}}. \qquad (1.11)$$

1.4.5.   *Non-ideal diode behaviour*

The 'ideal' diode behaviour of Eq. 1.5 is seldom seen. It is common for the dark current to depend more weakly on bias. The actual dependence on V is quantified by an ideality factor, m and the current–voltage characteristic given by the non-ideal diode equation,

$$J = J_{sc} - J_o(e^{qV/mk_B T} - 1) \qquad (1.12)$$

m typically lies between 1 and 2. The reasons for non-ideal behaviour will be discussed in later chapters.

1.5.   Summary

A photovoltaic cell consists of a light absorbing material which is connected to an external circuit in an asymmetric manner. Charge carriers are generated in the material by the absorption of photons of light, and are driven towards one or other of the contacts by the built-in spatial asymmetry. This light driven charge separation establishes a photovoltage at open circuit, and generates a photocurrent at short circuit. When a load is connected to the external circuit, the cell produces both current and voltage and can do electrical work.

The size of the current generated by the cell in short circuit depends upon the intensity and the energy spectrum of the incident light. Photocurrent is related to incident spectrum by the quantum efficiency of the cell, which is the probability of generating an electron per incident photon as a function of photon energy. When a load is present, a potential difference is created between the terminals of the cell and this drives a current, usually called the dark current, in the opposite direction to the photocurrent. As the load resistance is increased, the potential difference increases and the net current decreases until the photocurrent and dark current exactly cancel out. The potential difference at this point is called the open circuit voltage. At some point before V_{oc} is reached, the current–voltage product is maximum. This is the maximum power point and the cell should be operated with a load resistance which corresponds to this point.

The solar cell can be modelled as a current generator in parallel with an ideal diode, and the current–voltage characteristic given by the ideal diode equation, Eq. 1.5. In real cells, the behaviour is degraded by the presence of series and parallel resistances.

References

D. Anderson, *Clean Electricity from Photovoltaics*, eds. M.D. Archer and R.D. Hill (London: Imperial College Press, 2001).

M.A. Green, "Photovoltaics: Coming of age", *Conf. Record 21st IEEE Photovoltaic Specialists Conf.*, 1–7 (1990).

E. Lorenzo, *Solar Electricity: Engineering of Photovoltaic Systems* (Progensa, 1994).

T. Markvart, *Solar Electricity* (Wiley, 2000).

J.N. Shive, *Semiconductor Devices* (Van Nostrand, 1959).

C.A. Vincent, *Modern Batteries* (Arnold, 1997).

M. Wolf, "Historical development of solar cells", *Proc. 25th Power Sources Symposium*, 1972. In *Solar Cells*, ed. C.E. Backus (IEEE Press, 1976).

Chapter 2

Photons In, Electrons Out: Basic Principles of PV

2.1. Introduction

In Chapter 1 the solar cell was introduced and its performance character-istics, in response to applied bias and light, were defined. In this chapter we address some of the thermodynamic aspects of photovoltaic solar en-ergy conversion. The chapter is organised as follows: first the radiant power available from the sun is defined; the photovoltaic cell is distinguished from other types of solar energy converter and the question of how much elec-trical work can be extracted is addressed; the principle of detailed balance is introduced and used to calculate the performance characteristics of an ideal photovoltaic energy converter. We shall see that efficiency depends on the band gap of the absorbing material and the incident spectrum. Finally, the properties which are desirable for high efficiency in real photovoltaic materials and devices are discussed.

2.2. The Solar Resource

The sun emits light with a range of wavelengths, spanning the ultraviolet, visible and infrared sections of the electromagnetic spectrum. Figure 2.1 shows the amount of radiant energy received from the sun per unit area per unit time — the solar irradiance — as a function of wavelength at a point outside the Earth's atmosphere. Solar irradiance is greatest at visible wavelengths, 300–800 nm, peaking in the blue–green.

This extraterrestrial spectrum resembles the spectrum of a *black body* at 5760 K. A black body emits quanta of radiation — photons — with a distribution of energies determined by its characteristic temperature, T_s. At a point **s** on the surface of the black body the number of photons with energy in the range E to $E + dE$ emitted through unit area per unit solid

17

angle per unit time, the *spectral photon flux* $\beta_s(E, \mathbf{s}, \theta, \phi)$, is given by

$$\beta_s(E, \mathbf{s}, \theta, \phi) d\Omega.dSdE = \frac{2}{h^3 c^2} \left(\frac{E^2}{e^{E/k_B T_s} - 1} \right) d\Omega.dSdE \qquad (2.1)$$

where dS the element of surface area around $\mathbf{s}$ and $d\Omega$ the unit of solid angle around the direction of emission of the light (θ, ϕ). The flux issued normal to the surface is given by the component of β_s integrated over solid angle and resolved along $d\mathbf{S}$,

$$b_s(E, \mathbf{s}) dSdE = \int_\Omega \beta_s(E, \mathbf{s}, \theta, \phi) \cdot \cos\theta d\Omega.dSdE$$

$$= \frac{2F_s}{h^3 c^2} \left(\frac{E^2}{e^{E/k_B T_s} - 1} \right) dSdE \qquad (2.2)$$

where F_s is a geometrical factor which arises from integrating over the relevant angular range. Just at the surface of the black body this range is a hemisphere and $F_s = \pi$. Away from the surface, the angular range is reduced and

$$F_s = \pi \sin^2 \theta_{\text{sun}} \qquad (2.3)$$

where θ_{sun} is the half angle subtended by the radiating body to the point where the flux is measured. For the sun as seen from the earth, $\theta_s = 0.26°$ so that F_s is reduced by a factor of 4.6×10^4 to $2.16 \times 10^{-5}\pi$. If the temperature at all points $\mathbf{s}$ on the surface of the black body is the same, then the argument $\mathbf{s}$ can be dropped from b_s, and Eq. 2.2 can be written

$$b_s(E) = \frac{2F_s}{h^3 c^2} \left(\frac{E^2}{e^{E/k_B T_s} - 1} \right). \qquad (2.2)$$

In the remaining sections of this chapter we will use b_s and F_s to represent the spectral photon flux and geometrical factor for the sun.

Box 2.1. The angular resolved photon flux density, β, is the number of photons of given energy passing through unit area in unit time, per unit solid angle. It is defined on an element of surface area, and its direction is defined by the angle to the surface normal, θ, and an azimuthal angle, ϕ, projected on the plane of the surface element. In structures with planar symmetry it is sufficient to know the photon flux density resolved along the normal to the surface, b. b is obtained by integrating the components of β normal to the surface over solid angle.

The emitted energy flux density or *irradiance*, $L(E)$, is related to the photon flux density through

$$L(E) = Eb_{\mathrm{s}}(E). \tag{2.4}$$

Integrating Eq. 2.4 over E gives the total emitted power density, $\sigma_{\mathrm{S}} T_{\mathrm{S}}$, where σ_{S} is Stefan's constant,

$$\sigma_{\mathrm{S}} = \frac{2\pi^5 k^4}{15 c^2 h^3}.$$

At the sun's surface this is a power density of 62 MW m^{-2}. At a point just outside the Earth's atmosphere the solar flux is reduced (on account of the reduced angular range of the sun) and the solar power density is reduced to 1353 W m^{-2}. In Fig. 2.1 the extraterrestrial solar spectrum is compared with the spectrum of a 5760 K black body, reduced by the factor 4.6×10^4. The higher T_{s}, the higher the average energy of the emitted radiation. A black body at the temperature of the Earth, $T_{\mathrm{s}} = 300$ K, emits most strongly in the far infrared and its radiation cannot be seen. For the sun, with $T_{\mathrm{s}} = 5760$ K the emission is strongest at visible wavelengths. A hotter sun would emit light that appears blue to us, with a spectrum shifted to shorter wavelengths on Fig. 2.1, and a cooler sun would appear red.

On passing through the atmosphere, light is absorbed and scattered by various atmospheric constituents, so that the spectrum reaching the Earth's

The Physics of Solar Cells

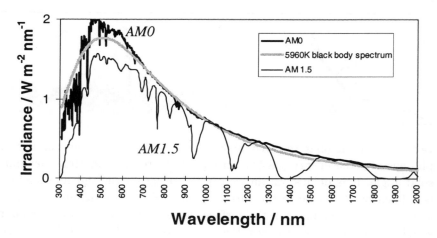

Fig. 2.1. Extra-terrestrial (Air Mass 0) solar spectrum (black line) compared with the 5760 K black body spectrum reduced by the factor 4.6×10^4 (thick grey line) and with the standard terrestrial (Air Mass 1.5) spectrum (thin grey line).

surface is both attenuated and changed in shape. Light of wavelengths less than 300 nm is filtered out by atomic and molecular oxygen, ozone, and nitrogen. Water and CO_2 absorb mainly in the infrared and are responsible for the dips in the absorption spectrum at 900, 1100, 1400 and 1900 nm (H_2O) and at 1800 and 2600 nm (CO_2). Attenuation by the atmosphere is quantified by the 'Air Mass' factor, n_{AirMass} defined as follows

$$n_{\mathrm{AirMass}} = \frac{\textbf{optical path length to Sun}}{\textbf{optical path length if Sun directly overhead}}$$
$$= \operatorname{cosec} \gamma_s . \qquad (2.5)$$

where γ_s is the angle of elevation of the sun, as shown in Fig. 2.2. The Air Mass n_{AirMass} spectrum is an extraterrestrial solar spectrum attenuated by n_{AirMass} thicknesses of an Earth atmosphere of standard thickness and composition.

The standard spectrum for temperature latitudes is *Air Mass 1.5*, or AM1.5, corresponding to the sun being at an angle of elevation of 42°. This atmospheric thickness should attenuate the solar spectrum to a mean irradiance of around 900 W m^{-2}. However, for convenience, the standard terrestrial solar spectrum is defined as the AM1.5 spectrum normalised so that the integrated irradiance is 1000 W m^{-2}. Actual irradiances clearly vary on account of seasonal and daily variations in the position of the sun

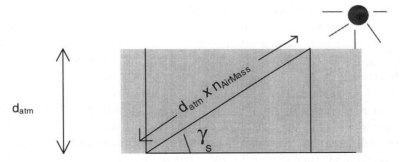

Fig. 2.2. If the atmosphere has thickness d_{atm}, then when the sun is at an angle of elevation γ_s, light from the sun has to travel through a distance $d_{atm} \times \mathrm{cosec}\, \gamma_s$ through the atmosphere to an observer on the Earth's surface. The optical depth of the atmosphere is increased by a factor $n_{AirMass} = \mathrm{cosec}\, \gamma_s$ compared to when the sun is directly overhead.

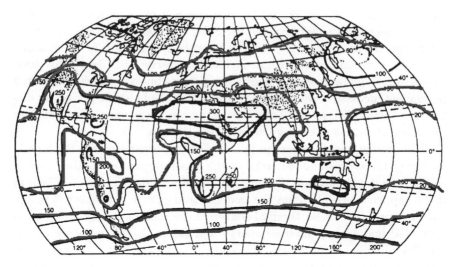

Fig. 2.3. Global distribution of annual average solar irradiance. The values on the irradiance contours are given in $\mathrm{W\ m^{-2}}$.

and orientation of the Earth and condition of the sky. Averaged global irradiances vary from less than 100 $\mathrm{W\ m^{-2}}$ at high latitudes to over 300 $\mathrm{W\ m^{-2}}$ in the sunniest places (usually, desert area in continental interiors), as shown in Fig. 2.3. (Solar radiation and spectral variations are discussed by [Gottschalg, 2001].)

For efficient solar collection, the solar collector should be directly facing the sun. However, variations in the position of the sun mean that any flat plate collector in a fixed position will face the sun only part of the time. Tracking systems can be used to follow the sun but these increase the cost.

Scattering of light by the atmosphere means a fraction of the light is *diffuse*, *i.e.*, incident from all angles rather than direct from the sun. This fraction is around 15% on average, but larger at higher latitudes, and in regions where there is a significant amount of cloud cover. Diffuse light presents different challenges for photovoltaic conversion. Since the light rays are not parallel, they cannot be refracted or concentrated. Materials with rough surfaces are relatively better suited for diffuse light than perfectly flat surfaces and are less sensitive to movements of the sun.

2.3. Types of Solar Energy Converter

The photovoltaic device should be distinguished from both solar thermal and photochemical energy converters. Solar thermal energy conversion results from the heat exchange between a hot body (the sun) and a cool one (the solar thermal device). Photochemical conversion is, like photovoltaic conversion, a quantum energy conversion process but one which results in a permanent increase in chemical potential rather than electric power. To distinguish these different types of solar energy converter, we need to consider the different modes of energy transfer from the sun.

The radiant energy absorbed by a device can either increase the kinetic energy of the atoms and electrons in the absorbing material (the internal energy), or it can increase the potential energy of the electrons. Which of these happens depends upon the material and how it is connected to the outside world. In a solar thermal converter the radiant energy absorbed is converted mainly into internal energy and raises the temperature of the cell. The difference in temperature relative to the ambient means that the solar converter can operate as a heat engine and do work, for instance by driving a steam turbine to generate electric power. Solar thermal converters utilise the full range of solar wavelengths, including the infrared, and are designed to heat up easily. They are thermally insulated from the ambient to make the working temperature difference as large as possible.

A photovoltaic converter, on the other hand, is designed to convert the incident solar energy mainly into electrochemical potential energy. Absorption of a photon in matter causes the promotion of an electron to a state of higher energy (an *excited state*). For the extra electronic energy to be extracted, the excited state should be separated from the ground state by an energy gap which is large compared to k_BT, where k_B is Boltzmann's constant. Therefore the material should contain two or more energy levels, or bands, which are separated by more than k_BT. In Chapter 3 we will see that a semiconductor is a very good example of such a system. The separation of the energy bands, or *band gap*, serves to maintain the excited electrons at the higher energy for a long time compared to the thermal relaxation time, so that they may be collected. Electrons in each of the different bands relax to form a local thermal equilibrium, called a *quasi thermal equilibrium*, with a different chemical potential, or, *quasi Fermi level*. In a two band system, the increase in electrochemical potential energy is given by the Gibbs free energy, $N\Delta\mu$, where N is the number of electrons promoted and $\Delta\mu$ the difference in the chemical potentials between the excited population and the ground state population. The difference in $\Delta\mu$ which results from the absorption of light is sometimes called the chemical potential of radiation. In equilibrium, $\Delta\mu = 0$. Extraction of electrochemical potential energy from light in this way is most effective when the ground state is full initially and the excited state is empty.

Unlike the solar thermal converter, the photovoltaic converter extracts solar energy only from those photons with energy sufficient to bridge the band gap. Since these mainly increase the electrochemical potential energy the increase in internal energy is much less. In practice, increased temperature can decrease the efficiency of photovoltaic conversion and so photovoltaic cells are usually designed to be in good thermal contact with the ambient.

To complete the photovoltaic conversion process, the excited electrons must be extracted and collected. This requires a mechanism for *charge separation*. Some intrinsic asymmetry is needed to drive the excited electrons away from their point of creation. (In general, charge separation involves positive holes and/or ions as well as electrons. We describe the process in terms of electrons for simplicity.) This can be provided by selective contacts such that carriers with raised μ (excited state) are collected at one contact and those with low μ (ground state) at the other. The difference in chemical potential between the contacts, $\Delta\mu$, then provides a potential difference between the terminals of the cell. Once separated, the charges should be allowed to travel without loss to an external circuit and do electrical work.

Photovoltaic conversion is similar to photochemical energy conversion (*e.g.* in photosynthesis), in that radiant energy produces an increase in electronic potential energy, rather than heat. In the case of photosynthesis

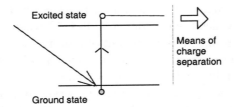

Fig. 2.4.   Excitation and charge separation. After an electron is promoted to a higher energy level by absorption of a photon of sufficient energy, it must be pulled away from the point of promotion by some mechanism for charge separation. The driving force for charge separation prevents the relaxation of the system to its initial state.

the excited electron population drives a chemical reaction, the conversion of CO_2 and water into carbohydrate, rather than driving an electric current. But in either case the solar energy results in a net flux of electronic potential energy constituting work. The different modes of solar energy conversion are explained in detail by de Vos [de Vos, 1992].

In the following sections we will calculate the amount of work available from a photovoltaic device.

2.4.   Detailed Balance

One of the fundamental physical limitations on the performance of a photovoltaic cell arises from the principle of detailed balance. As well as absorbing solar radiation the solar energy converter exchanges *thermal* radiation with its surroundings. Both the cell and the surrounding environment radiate long wavelength, thermal, photons on account of their finite temperature. The rate of emission of photons by the cell must be matched by the rate of photon absorption, so that in the steady state the concentration of electrons in the material remains constant.

2.4.1.   *In equilibrium*

First we consider the cell in the dark, in thermal equilibrium with the ambient. Assuming that the ambient radiates like a black body at a temperature T_a, then, according to Eq. 2.1, it produces a spectral photon flux at a point s on the surface of the solar cell of

$$\beta_a(E, \mathbf{s}, \theta, \phi)d\Omega.d\mathbf{S}dE = \frac{2}{h^3c^2}\left(\frac{E^2}{e^{E/k_BT_a} - 1}\right)d\Omega.d\mathbf{S}dE\,.$$

Integrating over directions, we obtain the incident flux of thermal photons normal to the surface of a flat plate solar cell

$$b_a(E) = \frac{2F_a}{h^3 c^2} \left(\frac{E^2}{e^{E/k_B T_a} - 1} \right) \qquad (2.6)$$

where the geometrical factor $F_a = \pi$, assuming that ambient radiation is received over a hemisphere. The equivalent *current density* absorbed from the ambient is

$$j_{abs}(E) = q(1 - R(E))a(E)b_a(E) \qquad (2.7)$$

where $a(E)$ is the probability of absorption of a photon of energy E and $R(E)$ is the probability of photon reflection. $j_{abs}(E)$ is the electron current density equivalent to the absorbed photon flux if each photon of energy E generates one electron. $a(E)$ is known as the *absorbance* or *absorptivity*, and is determined by the absorption coefficient of the material and by the optical path length through the device.

To obtain the total equivalent current for photon absorption, Eq. 2.7 should be integrated over the surface of the solar collector. The result depends on the interface at the rear surface. If the rear surface contacts the air, then both sides contribute equally, and the equivalent current is $2qA(1 - R(E))a(E)b_a(E)$ for a collector of area A. If the rear surface is in contact with a material of higher refractive index, n_s, the rate of photon absorption is enhanced by n_s^2 over that surface, and the result is $q(1 + n_s^2)A(1 - R(E))a(E)b_a(E)$. In the case of a perfect reflector (which is capable of reflecting thermal photons) at the rear surface, the equivalent current for absorbed thermal photons is only $qA(1 - R(E))a(E)b_a(E)$. In this case the areas for thermal photon and solar photon absorption are the same, and the device efficiency is the greatest. In the following analysis we assume that this is the case.

As well as absorbing thermal photons, the cell *emits* thermal photons by spontaneous emission. Spontaneous emission is the conversion into a photon of the potential energy released when an excited electron relaxes to its ground state (Fig. 2.5). (Stimulated emission, discussed in Chapter 4, can be neglected since the solar cell operates in a limit where the excited state is almost empty.) This emission is necessary to maintain a steady state.

A cell in thermal equilibrium with its surroundings, *i.e.*, receiving no radiation other than from the ambient, has temperature T_a and emits thermal radiation characteristic of that temperature. If ε is the *emissivity* (or probability of emission of a photon of energy E) the equivalent current

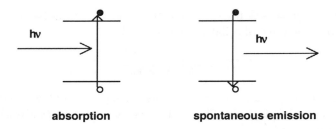

absorption   **spontaneous emission**

Fig. 2.5.   Absorption and spontaneous emission. In spontaneous emission, also known as radiative recombination, the electron relaxes from excited state to ground state giving out its extra potential energy as a photon of light.

density for photon emission through the surface of the cell is given by

$$j_{\text{rad}}(E) = q(1 - R(E))\varepsilon(E)b_{\text{a}}(E) \,. \tag{2.8}$$

In order to maintain a steady state, the current densities j_{abs} (Eq. 2.7) and j_{rad} (Eq. 2.8) must balance and therefore

$$\varepsilon(E) = a(E) \,. \tag{2.9}$$

This is a result of *detailed balance*: In quantum mechanical terms, it results from the fact that the matrix element for optical transitions from ground to excited state and from excited to ground state must be identical.

2.4.2.   *Under illumination*

Under illumination by a solar photon flux $b_{\text{s}}(E)$ (Eq. 2.2), the cell absorbs solar photons of energy E at a rate

$$(1 - R(E))a(E)b_{\text{s}}(E) \,.$$

The equivalent current density for photon absorption includes a contribution from thermal photons, hence

$$j_{\text{abs}}(E) = q(1 - R(E))a(E)\left(b_{\text{s}}(E) + \left(1 - \frac{F_{\text{s}}}{F_{\text{e}}}\right)b_{\text{a}}(E)\right) \tag{2.10}$$

where the coefficient of b_{a} is introduced to allow for the fraction of the incident ambient flux which has been replaced by solar radiation.

As a result of illumination, part of the electron population has raised electrochemical potential energy, and the system develops a chemical potential $\Delta\mu > 0$. In these conditions spontaneous emission is increased and the rate of emission depends upon $\Delta\mu$. This makes sense since when more electrons are at raised energy, relaxation events are more frequent. According to a generalised form of Planck's radiation law, the spectral photon flux emitted from a body of temperature T_C and chemical potential $\Delta\mu$ into a medium of refractive index n_s is given by

$$\beta(E, s, \theta, \phi) = \frac{2n_s^2}{h^3 c^2} \frac{E^2}{e^{(E-\Delta\mu)/k_B T_a} - 1} \tag{2.11}$$

per unit surface area and solid angle [Wuerfel, 1982; de Vos, 1992]. Integrating over the range of solid angle through which photons can escape $(0 \leq \theta \leq \theta_c)$ we obtain the photon flux emitted normal to the surface

$$b_e(E, \Delta\mu) = F_e \frac{2n_s^2}{h^3 c^2} \frac{E^2}{e^{(E-\Delta\mu)/k_B T_a} - 1} \tag{2.12}$$

where

$$F_e = \pi \sin^2 \theta_c = \pi \frac{n_0^2}{n_s^2} \tag{2.13}$$

and

$$\theta_c = \sin^{-1} \left(\frac{n_0}{n_s} \right)$$

by Snell's law, where n_0 is the refractive index of the surrounding medium.

At a surface with air, $n_0 = 1$, $F_e \times n_s^2 = F_a = \pi$ and

$$b_e(E, \Delta\mu) = \frac{2F_a}{h^3 c^2} \frac{E^2}{e^{(E-\Delta\mu)/k_B T_a} - 1} . \tag{2.14}$$

Note that this result is the same whether the integration is taken over internal or external solid angle: internally, n_s must be retained but the angular range is limited to θ_c, while externally $n_s = 1$ but the angular range is a hemisphere.

Now if ε is the probability of photon emission, the equivalent current density for photon emission is

$$j_{rad}(E) = q(1 - R(E))\varepsilon(E)b_e(E, \Delta\mu) . \tag{2.15}$$

It is easy to see that Eq. 2.15 reduces to Eq. 2.8 for the cell in equilibrium, where $a = \varepsilon$ and $\Delta\mu = 0$. It is not immediately obvious how $a(E)$ relates to $\varepsilon(E)$ for the cell with $\Delta\mu > 0$. However, it has been shown elsewhere [Araujo, 1994] from a generalised detailed balance argument that Eq. 2.9

still holds, provided that $\Delta\mu$ is constant through the device. That result will be used below without proof.

The *net* equivalent current density, from Eqs. 2.10 and 2.15 is,

$$j_{\mathrm{abs}}(E) - j_{\mathrm{rad}}(E)$$

$$= q(1 - R(E))a(E)\left(b_{\mathrm{s}}(E) + \left(1 + \frac{F_{\mathrm{s}}}{F_{\mathrm{a}}}\right)b_{\mathrm{a}}(E) - b_{\mathrm{e}}(E, \Delta\mu)\right). \quad (2.16)$$

This may be divided into contributions from *net* absorption (in excess to that at equilibrium),

$$j_{\mathrm{abs(net)}}(E) = q(1 - R(E))a(E)\left(b_{\mathrm{s}}(E) - \frac{F_{\mathrm{s}}}{F_{\mathrm{e}}}b_{\mathrm{a}}(E)\right) \quad (2.17)$$

and the *net* emission, or *radiative recombination* current density

$$j_{\mathrm{abs(net)}}(E) = q(1 - R(E))a(E)(b_{\mathrm{e}}(E, \Delta\mu) - b_{\mathrm{e}}(E, 0)), \quad (2.18)$$

noting that $b_{\mathrm{a}}(E) = b_{\mathrm{e}}(E, 0)$. This radiative recombination is an unavoidable loss which means that absorbed solar radiant energy can never be fully utilised by the solar cell. Radiative recombination is discussed further in Chapter 4.

2.5. Work Available from a Photovoltaic Device

Now we have enough information to calculate the absolute limiting efficiency of a photovoltaic converter. We will consider a two band system for which the ground state (lower band) is initially full and the excited state (upper band) empty. The bands are separated by a band gap, E_{g}, so that light with $E < E_{\mathrm{g}}$ is not absorbed (see Fig. 2.6). We will assume that electrons in each band are in quasi thermal equilibrium at the ambient temperature T_{a} and the chemical potential for that band, μ_{i}.

2.5.1. *Photocurrent*

Photocurrent is due to the net absorbed flux due to the sun, Eq. 2.17. Since the angular range of the sun is so small compared to the ambient, the second term in Eq. 2.17 is usually neglected. If each electron has a probability, $\eta_{\mathrm{c}}(E)$, of being collected, we obtain the photocurrent density

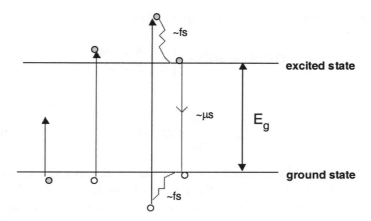

Fig. 2.6. Two band photoconverter. Photons with energy $E < E_g$ cannot promote an electron to the excited state. Photons with $E \geq E_g$ can raise the electron but any excess energy is quickly lost as heat as the carriers relax to the band edges. An absorbed photon with $E \gg E_g$ achieves the same result as a photon with $E = E_g$. For this reason it is the incident photon flux and not the photon energy density which determines the photogeneration. Once excited the electrons remain in the excited state for a relatively long time.

at short circuit by integrating j_{abs} over photon energies

$$J_{sc} = q \int_0^\infty \eta_c(E)(1 - R(E))a(E)b_s(E)dE \,. \qquad (2.19)$$

This is identical to Eq. 1.1 with the quantum efficiency $QE(E)$ given by the product of the collection and absorption efficiencies.

$$J_{sc} = q \int_0^\infty QE(E)b_s(E)dE \qquad (1.1)$$

For the case of the most efficient solar cell we will suppose that we have a perfectly absorbing, non-reflecting material so that that *all* incident photons of energy $E > E_g$ are absorbed to promote exactly one electron to the upper band. We further suppose perfect charge separation so that all electrons which survive radiative recombination are collected by the negative terminal of the cell and delivered to the external circuit (*i.e.* $\eta_c(E) = 1$). This gives the maximum photocurrent for that band gap, assuming that multiple carrier generation — the promotion of *more* than one electron by an absorbed photon — does not happen. Then

$$QE(E) = a(E) = \begin{cases} 1 & E \geq E_g \\ 0 & E < E_g \end{cases} \qquad (2.20)$$

and

$$J_{sc} = q \int_{E_g}^{\infty} b_s(E)dE . \qquad (2.21)$$

Photocurrent is then a function *only* of the band gap and the incident spectrum. Clearly, the lower E_g, the greater will be J_{sc}. It is also clear from Eq. 2.21 that it is necessary to define the spectrum for any statement of efficiency.

2.5.2. *Dark current*

Dark current is the current that flows through the photovoltaic device when a bias is applied in the dark. We will suppose that in the ideal cell material no carriers are lost through non-radiative recombination, for example at defects within the material. The only loss process considered is the unavoidable radiative relaxation of electrons through spontaneous emission, described above. The dark current density due to this process is given by integrating j_{rad} over photon energy and, for a flat plate cell with perfect rear reflector, is given by

$$J_{rad}(\Delta\mu) = q \int (1 - R(E))a(E)(b_e(E, \Delta\mu) - b_e(E, 0))dE , \qquad (2.22)$$

assuming that $\Delta\mu$ is constant over the surface of the cell and using the detailed balance result, $a(E) = \varepsilon(E)$. In ideal material with lossless carrier transport $\Delta\mu$ can be further assumed constant *everywhere* and equal to q times the applied bias V [Araujo, 1994]. Then, assuming that dark current and photocurrent can be added, as in Eq. 1.3

$$J(V) = J_{sc} - J_{dark}(V) ,$$

we obtain for the net cell current density,

$$J(V) = q \int_0^{\infty} (1 - R(E))a(E)\{b_s(E) - (b_e(E, qV) - b_e(E, 0))\}dE . \qquad (2.23)$$

For the special case of the step-like absorption function (Eq. 2.20),

$$J(V) = q \int_{E_g}^{\infty} \{b_s(E) - (b_e(E, qV) - b_e(E, 0))\} dE, \qquad (2.24)$$

$J(V)$ is strongly bias dependent through the exponential term in Eq. 2.12 and has the approximate form

$$J(V) = J_{sc} - J_0(e^{qV/k_B T} - 1)$$

where J_0 is a (temperature dependent) constant for the particular material. This resembles the ideal diode Eq. 1.4.

The net electron current is thus due to the difference between the two photon flux densities: the absorbed flux, which is distributed over a wide range of photon energies above the threshold E_g, and the emitted flux, which is concentrated on photon energies near E_g. As V increases, the emitted flux increases and the net current decreases. At the open circuit voltage V_{oc} the total emitted flux exactly balances the total absorbed flux and the net current is zero. If V is increased still further, the emitted flux exceeds the absorbed and the cell begins to act like a light emitting device, giving out light in return for the applied electrical potential energy. Note that V_{oc} must always be less than $\frac{E_g}{q}$. The spectral fluxes leading to these regimes are illustrated in Fig. 2.7(a), while Fig. 2.7(b) illustrates the resulting $J(V)$ curves.

2.5.3. *Limiting efficiency*

To calculate the power conversion efficiency we need to calculate the incident and extracted *power* from the photon fluxes. The incident power density is obtained simply by integrating the incident irradiance (Eq. 2.4) over photon energy,

$$P_s = \int_0^{\infty} E b_s(E_s) dE. \qquad (2.25)$$

For the output power we need to know the electrical potential energy of the extracted photo-electrons. For the ideal photoconverter it is assumed that no potential is lost through resistances anywhere in the circuit. Therefore all collected electrons should have $\Delta\mu$ of electrical potential energy and deliver $\Delta\mu$ of work to the external circuit. Since $\Delta\mu = qV$ we have for the

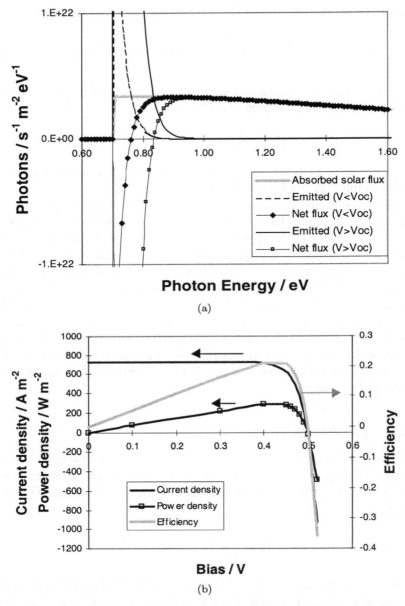

Fig. 2.7. (a) Absorbed ($b_s(E)$), emitted ($b_e(E, qV)$) and net ($= b_s - b_e$) spectral photon flux for a biased cell of $E_g = 0.7$ V at 300 K illuminated by a black body sun at 5760 K. (b) Current density, power density and efficiency of the device in (a) as a function of V. The current is calculated from q times the integrated net photon flux.

extracted power density from Eq. 1.6

$$P = VJ(V)$$

with $J(V)$ given by Eq. 2.24 above. The power conversion efficiency is

$$\eta = \frac{VJ(V)}{P_s}. \qquad (2.26)$$

Maximum efficiency is achieved when

$$\frac{d}{dV}(J(V)V) = 0. \qquad (2.27)$$

The bias at which this occurs is the maximum power bias, V_m introduced in Chapter 1. In Fig. 2.7(b) the output power density for a 0.7 eV band gap photoconverter in a black body sun is plotted as a function of bias. At the maximum, $V_m = 0.45$ V, the power conversion efficiency is around 20%.

2.5.4. *Effect of band gap*

Given all of the assumptions made above, the power conversion efficiency of the ideal two band photoconverter is a function only of E_g and the incident spectrum. If the incident spectrum is fixed, then η depends only on the band gap. Intuitively we can see that very small and very large band gaps will lead to poor photoconverters: in the first case because the working value of V is too small, (V_m, like V_{oc}, is always less than E_g) and in the second because the photocurrent is too small. For any spectrum there is an optimum band gap at which η has a maximum. Figure 2.8 shows the variation of η with E_g calculated in this way for the standard AM1.5 solar

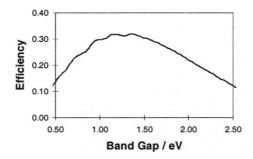

Fig. 2.8.   Calculated limiting efficiency for a single band gap solar cell in AM 1.5.

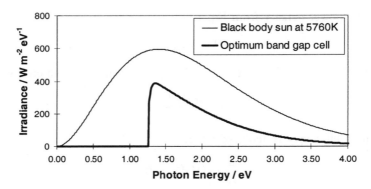

Fig. 2.9. Power spectrum of a black body sun at 5760 K, and power available to the optimum band gap cell.

spectrum. It has a maximum of about 33% at an E_g of around 1.4 eV. Optimising the performance of the ideal single band gap photoconverter is therefore a matter of choosing the right material.

In Fig. 2.9 the available power spectrum for an optimum band gap cell at maximum power point is compared with the incident power from a black body sun. Clearly, no photons with energy less than E_g contribute to the available power. Photons of $E > E_g$ are absorbed but deliver only $\Delta\mu(= qV_m)$ of electrical energy to the load, so only $\Delta\mu/E$ of their power is available. The figure shows how this fraction falls as E increases. Even at $E = E_g$ only a fraction $\Delta\mu/E_g$ of the incident power is available, since $qV_m < E_g$.

2.5.5. *Effect of spectrum on efficiency*

To model the influences of spectrum on limiting efficiency, it is convenient to use a black body spectrum at T_s as the illuminating source. The spectrum of a 5760 K black body with the angular width of the sun is a good model of the extra-terrestrial (Air Mass 0) spectrum and predicts a limiting efficiency of around 31% at a band gap of 1.3 eV [Araujo, 1994], somewhat lower than the maximum efficiency in AM1.5.

If the spectrum is shifted to the red, by reducing the temperature of the source, the optimum band gap and the limiting efficiency are both reduced. Clearly, in the limit where $T_s = T_a$ the cell is in equilibrium with the source and there is no net photoconversion. On the other hand, if the temperature of the source is increased relative to the cell, so is the photoconversion efficiency. In the limit where $T_a \to 0$, the radiative current vanishes and bias has no effect on the net photocurrent. Then the optimum operating bias is $V = E_g/q$ (anything higher is physically unreasonable) and if all carriers are collected with $\Delta\mu = qV$ then the maximum efficiency is given by

$$\eta = \frac{E_g \int_{E_g}^{\infty} b_s(E)dE}{\int_0^{\infty} E b_s(E)dE} .$$

This has a maximum of around 44% at a band gap of 2.2 eV for a 6000 K black body sun, increasing to higher values and higher band gaps for hotter suns. This limit was reported by Shockley and Queisser [Shockley, 1961] as the ultimate efficiency of the solar cell. In practice the cooling of the cell below the ambient requires an input of energy which reduces the net efficiency.

Another way of improving the efficiency through the spectrum is to alter the angular width of the sun. Recall from Eq. 2.2 that the solar flux contains a factor F_s which represents the solid angle subtended by the sun. If this angle is increased by *concentrating* the light, the net photocurrent will increase and the first term (absorbed flux) in the integrand in Eq. 2.24 will increase relative to the second (emitted flux). One way of looking at this is to consider that while the cell emits radiation in all directions, it absorbs sunlight only from a small angular range. Increasing the angular range improves the balance, as does restricting the angular range for emission. This will be considered in more detail in Chapter 9. Optimising the power density then yields a new $\eta(E_g)$ curve with a higher maximum at a smaller band gap. For light which is concentrated by a factor of 1000, a limiting efficiency of about 37% at $E_g = 1.1$ eV is predicted [Henry, 1980]. For a concentration factor of 4.6×10^4 (the maximum) η is over 40%. However, these estimates ignore the practical effect that under high concentrations the cell will be heated, and emit more strongly.

2.6. Requirements for the Ideal Photoconverter

In the above we made the following assumptions:

- that our photovoltaic material has an energy gap which separates states which are normally full from states which are normally empty;
- that all incident light with $E > E_g$ is absorbed;
- that each absorbed photon generates exactly one electron-hole pair;

- that excited charges do not recombine except radiatively, as required by detailed balance;
- that excited charges are completely separated;
- that charge is transported to the external circuit without loss.

Let's examine what these assumptions mean for real physical systems.

Energy gap

Many solid state and molecular materials satisfy the condition of the energy or band gap. The need for conductivity make semiconductors particularly suitable. With band gaps in the range 0.5–3 eV semiconductors can absorb visible photons to excite electrons across the band gap, where they may be collected. The III–V compound semiconductors gallium arsenide (GaAs) and indium phosphide (InP) have band gaps close to the optimum (1.42 eV and 1.35 eV, respectively, at 300 K) and are favoured for high efficiency cells. The most popular solar cell material, silicon, has a less favourable band gap (1.1 eV, maximum efficiency of 29%) but is cheap and abundant compared to these III–V materials. Other compound semiconductors, in particlular cadmium telluride (CdTe) and copper indium gallium diselenide ($CuInGaSe_2$) are being developed for thin film photovoltaics. Recent developments in semiconducting molecular materials indicate that organic semiconductors are promising materials for photovoltaic energy conversion in the future.

Light absorption

High absorption of light with $E > E_g$ is straightforward to achieve in principle. Increasing the thickness of the absorbing layer increases its optical depth, and for most semiconductors almost perfect absorption can be achieved with a layer a few tens or hundreds of microns thick. However, the requirements of high optical depth *and* perfect charge collection, make very high demands of material quality.

Charge separation

For a current to be delivered, the material should be contacted in such a way that the promoted electrons experience a spatial *asymmetry*, which drives them away from the point of promotion. This can be an electric field, or a gradient in electron density.

This asymmetry can be provided by preparing a *junction* at or beneath the surface. The junction may be an interface between two electronically different materials or between layers of the same material treated in different ways. It is normally large in area to maximise the amount of solar energy intercepted. For *efficient* photovoltaic conversion the junction quality is of central importance since electrons should lose as little as possible of their electrical potential energy while being pulled away. In practice preparing this large area junction successfully and without detriment to material quality is a challenge and limits the number of suitable materials.

Lossless transport

To conduct the charge to the external circuit the material should be a good electrical conductor. Perfect conduction means that carriers must not recombine with defects or impurities, and should not give up energy to the medium. There should be no resistive loss (no series resistance) or current leakage (parallel resistance). The material around the junction should be highly conducting and make good Ohmic contacts to the external circuit.

Mechanisms for excitation, charge separation and transport can be provided by the semiconductor *p–n* junction, which is the classical model of a solar cell. In this system charge separation is achieved by a charged junction between layers of semiconductor of different electronic properties: *i.e.*, the driving force which separates the charges is electrostatic. The *p–n* junction will be treated in detail in Chapter 6.

Optimum load resistance

Finally, the load resistance should be chosen to match the operating point of the cell. As we have seen above, individual solar cells tends to offer photovoltages of less than one volt which are often too small to be useful. For most applications, voltage is increased by connecting several cells in series into a module, and sometimes by connecting modules in series and parallel into a larger array. In practice the load resistance should be matched with the maximum power point of the array, rather than the cell.

As a consequence of the demands on the material, only a very small number of materials, all of them inorganic semiconductors, have been developed for photovoltaics. Only a few of the many potentially useful materials have the necessary technological history. The favourites are those developed for the microelectronics industry — silicon, gallium arsenide, amorphous silicon, some II–VI and other III–V compounds. It is only recently that materials have been developed *primarily* for their application in photovoltaics.

In terms of the above discussion, the main reasons why real solar cells do not achieve ideal performance are these:

• Incomplete absorption of the incident light. Photons are reflected from the front surface or from the contacts or pass through the cell without being absorbed. This reduces the photocurrent.

• Non-radiative recombination of photogenerated carriers. Excited charges are trapped at defect sites and subsequently recombine before being collected. This can occur at the surfaces where the defect density is higher, or near interfaces with another material, or near the junction. Recombination reduces both the photocurrent, through the probability of carrier collection, and the voltage, by increasing the dark current.

• Voltage drop due to series resistance between the point of photogeneration and the external circuit. This reduces the available power, as discussed in Chapter 1. It also means that $\Delta\mu \neq qV$.

In following chapters we shall see how far different designs and materials meet the demands of the ideal photovoltaic converter.

2.7. Summary

The sun emits radiant energy over a range of wavelengths, peaking in the visible. Its spectrum is similar to that of a black body at 5760 K, although it is influenced by atmospheric absorption and the position of the sun. The standard solar spectrum for photovoltaic calibration is the AM 1.5 spectrum.

A photovoltaic solar energy converter absorbs photons of radiant energy to excite electrons to a higher energy level, where they have increased electrochemical potential energy. In order for these excited electrons to be extracted as electrical power, the material must possess an energy gap or band gap. To calculate the absolute limiting efficiency of a photovoltaic energy converter, we use the principle of detailed balance. This allows for the fact that any body which absorbs light must also emit light. A photovoltaic device will emit more light when optically excited on account of the extra electrochemical potential energy of the electrons. This radiative recombination is the mechanism which ultimately limits the efficiency of a photovoltaic cell. The current delivered by the ideal photoconverter is due to the difference between the flux of photons absorbed from the sun and the flux of photons emitted by the excited device, while the voltage is due to the electrochemical potential energy of the excited electrons. From this we calculate the current–voltage characteristic of an ideal solar cell. The maximum efficiency depends upon the incident spectrum and the band gap, and for a standard solar spectrum it is around 33% at a band gap of 1.4 eV. For a real device to approach the limiting efficiency, it should have an optimum energy gap, strong light absorption, efficient charge separation and charge transport, and the load resistance should be optimised.

References

G.L. Araujo and A. Marti, "Absolute limiting effciencies for photovoltaic energy conversion", *Solar Energy Materials and Solar Cells* **33**, 213 (1994).

A. de Vos, *Endoreversible Thermodynamics of Solar Energy Conversion* (Oxford University Press, 1992).

R. Gottschalg, *The Solar Resource and the Fundamentals of Radiation for Renewable Energy Systems* (Sci-Notes, Oxford, 2001).

C.H. Henry, "Limiting efficiencies of ideal single and multiple energy gap terrestrial solar cells", *J. Appl. Phys.* **51**, 4494–4499 (1980).

W. Shockley and H.J. Queisser, "Detailed balance limit of efficiency of *p–n* junction solar cells", *J. Appl. Phys.* **32**, 510–519 (1961).

P. Wuerfel, "The chemical potential of radiation", *J. Phys.* **C15**, 3697 (1982).

Chapter 3

Electrons and Holes in Semiconductors

3.1. Introduction

In Chapter 2 we considered the requirements for photovoltaic energy conversion. We concluded that a suitable photovoltaic material should absorb visible light, possess a band gap between the initial, occupied states and the final, unoccupied states which are involved in photon absorption and be able to transport charges efficiently. The gap is necessary in order to make the extra potential energy which electrons gain from photon absorption available as electrical energy. All semiconducting and insulating solids possess an energy gap but only semiconductors are suitable for photovoltaics, because the band gap of insulators is too large to permit absorption of visible light. Most molecular solids possess an energy gap, but with the exception of some conjugated molecular materials, the charge transport is too inefficient to be useful for solar cells.

The band gap is important because it enables excited electrons to remain in higher energy levels for long enough to be exploited. If electrons were simply promoted through a continuum of energy levels as in a metal, for example, they would very quickly decay back down to their ground state through a series of intermediate levels. The abundance of empty levels at intermediate energy means that the probability of an excited electron being scattered to a lower energy state within the thermal energy of the original level, is high. At room temperature this 'thermalisation' of carriers to the band edge occurs in femtoseconds. When an electron is excited across a band gap, it quickly decays to the lowest available energy state in the conduction band (the conduction band edge) but the next stage — decay across the band gap to a vacant site in the valence band — is slow, as shown in Fig. 3.1.

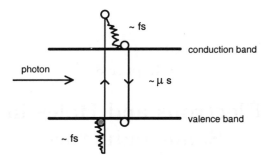

Fig. 3.1. Promotion of an electron from valence band to conduction band by a photon, thermalisation to the band edges, and recombination. Although thermalisation is very fast and occurs in femtoseconds, relaxation across the band gap is many orders of magnitude slower.

In practice, the great majority of experience with photovoltaic materials is based on a small number of semiconductor materials. In this chapter and the next we will examine some of the basic physical principles of semiconductors. We will focus on the electronic and optical properties of crystalline materials. We will show how the optical and electronic properties result from the crystal structure. In this chapter we introduce the concepts of density of states, electron distribution function, doping, quasi thermal equilibrium and the definition of electron and hole currents. In Chapter 4 we will treat the processes of charge carrier generation and recombination in semiconductors and show how to set up the semiconductor transport equations, which are key to the physics of photovoltaic devices.

3.2.   Basic Concepts

3.2.1.   *Bonds and bands in crystals*

When a pair of atoms are brought together into a molecule, their atomic orbitals combine to form pairs of molecular orbitals arranged slightly higher and slightly lower in energy than each original level. We say that the energy levels have *split*. When a very large number of atoms come together in a solid, each atomic orbital splits into a very large number of levels, so close together in energy that they effectively form a continuum, or *band*, of allowed levels. The bands due to different molecular orbitals may or may not overlap. The energy distribution of the bands depends upon the electronic properties of the atoms and the strength of the bonding between them.

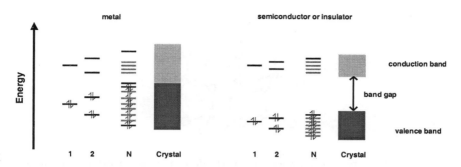

Fig. 3.2. As the number of atoms in a molecule or cluster increases, the atomic orbitals split into multiple levels, eventually coming together to form bands in the limit of many atoms. Overlapping (left) and non overlapping (right) bands represent a metal and a semiconductor, respectively.

Bands are occupied or not depending upon whether the original molecular orbitals were occupied. The highest occupied band, which contains the valence electrons, is normally called the *valence band* (VB). The lowest unoccupied band is called the *conduction band* (CB). If the valence band is partly full, or if it overlaps in energy with the lowest unoccupied band, the solid is a metal. In a metal, the availability of empty states at similar energies makes it easy for a valence electron to be excited, or *scattered*, into a neighbouring state. These electrons can readily act as transporters of heat or charge, and so the solid conducts heat and electric current.

If the valence band is completely full and separated from the next band by an energy gap, then the solid is a semiconductor or an insulator. The electrons in the valence band are all completely involved in bonding and cannot be easily removed. They require an energy equivalent to the band gap to be removed to the nearest available unoccupied level. These materials therefore do not conduct heat or electricity easily.

Semiconductors are distinguished, roughly, as the group of materials with a band gap in the range 0.5 to 3 eV. Semiconductors have a small conductivity in the dark because only a small number of valence electrons will have enough kinetic energy at room temperature to be excited across the band gap at room temperature. This *intrinsic* conductivity decreases with increasing band gap. *Insulators* are wider band gap materials whose conductivity is negligible at room temperature. Materials of band gap < 0.5 eV have a reasonably high conductivity and are usually known as *semimetals*.

When the solid forms a regular crystal, then the energies of the bands, or the *band structure* can be predicted exactly. Exactly which crystal structure

Fig. 3.3.   Structure of crystalline silicon. Each silicon atom is bonded to four others in a tetrahedral arrangement.

a solid will adopt depends upon the number of valence electrons and other factors. It will prefer a configuration that minimises the total energy. A band gap is likely to arise in a crystal structure where all valence electrons are used in bonding. For example, the silicon atom possesses four valence electrons in its outermost 3s and 3p atomic orbitals. If the atom could form bonds with four neighbours, each contributing one electron, then all valence electrons would be occupied in bonding. In crystalline silicon this is achieved by the hybridisation of the 3s and 3p orbitals into a set of four degenerate sp^3 orbitals, which are directed in space with tetrahedral symmetry, and allow the formation of four identical silicon-silicon bonds with neighbouring atoms (Fig. 3.3). When the crystalline solid is formed, the sp^3 orbitals split to form a pair of bands. The lower, bonding band is completely filled by the valence electrons and the upper, antibonding or conduction band, is completely empty in a perfect crystal at absolute zero. Some solids can exist in different phases. For instance, carbon can form either the highly insulating, wide band gap, diamond crystal structure where all four valence electrons are tied up in covalent bonds with neighbouring carbon atoms, or the semimetallic graphite structure where only three valence electrons are involved in directed bonds with neighbouring atoms while the remaining electron is loosely involved in bonding with another plane of carbon atoms and is relatively mobile.

3.2.2.   *Electrons, holes and conductivity*

At absolute zero temperature, a pure semiconductor is unable to conduct heat or electricity since all of its electrons are involved in bonding. As the temperature is raised, the electrons gain some kinetic energy from vibrations of the lattice and some are able to break free. The freed electrons have been excited into the conduction band and are able to travel and transport charge or energy. Meanwhile, the vacancies which they have left behind are able to move, and can also conduct (see Box 3.1). The higher the temperature, the greater the number of electrons and holes which are mobilised, and the higher the conductivity.

Box 3.1. Electrons and holes

When an electron is removed from a bond between atoms, a positively charged vacancy remains. This vacancy can be filled by another electron, most easily by electrons which are involved in neighbouring bonds. If this

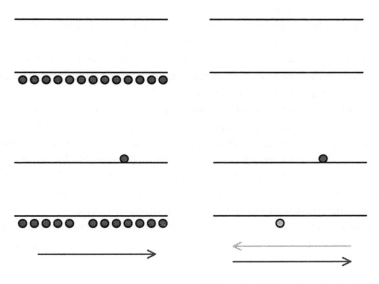

Valence and conduction electrons   Valence band holes and conduction electrons

Fig. 3.4. The valence band is represented, very crudely, by marbles in a tray. When the valence band is filled (top) no net electron motion can occur, despite an applied force, because there are no vacant orbitals for the electrons to be scattered into. Motion becomes possible when an electron is removed (bottom), freeing up some vacant orbitals. The nearly filled valence band is most easily characterised in terms of those orbitals which are not filled, called 'holes'. The dynamics of the hole are equivalent to those of a particle with the opposite charge, wavevector and energy to the missing electron. It responds to an applied field by moving in the opposite sense to an electron. The two situations depicted on the right are equivalent to those on the left, using holes to represent missing electrons.

happens, the vacancy moves to the neighbouring bond. The creation of holes in the valence band creates a means whereby charge can be transferred. In the presence of an electric field, a bonding, or *valence*, electron can respond to the field by moving into the hole. The vacancy which it leaves behind can be filled by another valence electron, and so on. The net movement of the valence electrons against the field is equivalent to the movement of a small number of positive *holes* in the direction of the field. Since there are many fewer valence holes than electrons it is much more convenient to think in terms of the motion of holes through the valence band. These positive holes can be characterised with a mobility and an effective mass, just like conduction electrons.

A semiconductor can be made to conduct in other ways: if the material is exposed to light of energy greater than the band gap, a photon can be absorbed by a valence electron to free it from the lattice and promote it into the conduction band. This creates a vacancy in the valence band. The free electron and hole created in this way are available to conduct electricity. This is called *photoconductivity*, the phenomenon which first drew attention to photovoltaic materials (see Chapter 1).

Alternatively, impurities with different numbers of valence electrons or different bond strengths can be added to the material. Electrons or holes may be freed more easily from these impurities than from the native atoms, thereby increasing the number of carriers normally available for conduction compared to the pure semiconductor. Deliberate addition of impurities to increase conductivity in this way is called *doping*. Both of these processes — photoconductivity and doping — will be discussed later.

In the next section we will consider what controls the energetic arrangement and the dynamics of the electrons and holes in a semiconductor.

3.3.   Electron States in Semiconductors

3.3.1.   *Band structure*

To locate the energy levels of an atom or a molecule, we need to solve Schrödinger's equation. In a solid we do the same thing, except that now we need to take account of an infinite array of atomic potentials, and not just the few which make up our molecule. In a crystalline material we are able to take advantage of the fact that the atomic potentials are arranged on an infinite periodic lattice. The *periodicity* of the lattice means that the probability distribution of the electrons must also be periodic. There is no physical reason why an electron should occupy a particular site within one unit cell rather than within any other. Because the lattice is *infinite*, the electrons should form delocalised states which extend throughout the crystal, just like an electron in free space. The sort of wavefunction which satisfies these conditions is the *Bloch* wavefunction. It is the product of a periodic part $u_{ik}(r)$, which possesses the periodicity of the lattice, and a plane wave part. The periodic part is related to one of the atomic orbitals involved in the crystal structure and inherits its symmetry. The plane wave modulates the wavefunction and is analogous to the wavefunction of an electron in free space. Each Bloch state is thus characterised by a crystal band i and a wavevector $\mathbf{k}$,

$$\psi(\mathbf{k}, \mathbf{r}) = u_{ik}(\mathbf{r})e^{i\mathbf{k}\cdot\mathbf{r}}. \tag{3.1}$$

In quantum mechanical terms, $\mathbf{k}$ is a 'good quantum number'. For each original atomic orbital i there is a continuous set of solutions with different $\mathbf{k}$ which make up the wavefunctions in the ith crystal band. The $u_{ik}(\mathbf{r})$ and eigenenergies $E(\mathbf{k})$ are, in general, found by solving the Schrödinger equation for each band i, for each $\mathbf{k}$. The map of energies E against wavevector $\mathbf{k}$ is called the crystal *band structure* (see Box 3.2). The $u_{ik}(\mathbf{r})$ are usually rapidly varying with distance and weakly dependent on $\mathbf{k}$. In a commonly used approximation, called the *effective mass approximation*, the $\mathbf{k}$ dependence of the $u_{ik}(\mathbf{r})$ is neglected. Then all of the dynamic information is contained in the plane wave part of the wavefunction, and the $u_i(\mathbf{r})$ come in through effective parameters.

Box 3.2. Band structure diagrams

$E(\mathbf{k})$ is usually plotted against $|\mathbf{k}|$ for the two or three most important directions in the crystal. Directions are labelled with Miller indices $\langle hkl \rangle$ where (h,k,l) is a vector parallel to $\mathbf{k}$, measured in unit cell widths along the natural axes of the crystal. For instance, the direction $\langle 100 \rangle$ represents the set of $\mathbf{k}$ parallel to the first of the natural axes, say the x axis. In a cubic system, $\langle 100 \rangle$ also represents the directions parallel to the y and z axes, because these three directions are indistinguishable in cubic symmetry.

k is varied from 0 to π/a along the chosen direction, where a is the spacing of the planes of atoms in that direction. For $k > \pi/a$, the energy spectrum repeats itself. This is because in a periodic structure with period a, wavevectors which differ by multiples of $2\pi/a$ cannot be distinguished.

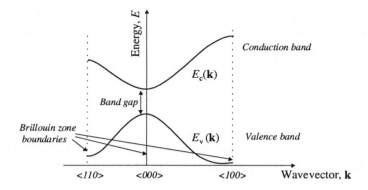

Fig. 3.5.   Schematic band structure diagram of a direct gap semiconductor.

For most crystals, for $k < 0$ the function $E(\mathbf{k})$ is identical to $E(-\mathbf{k})$ since states with positive and negative k are degenerate. Therefore, all the information about $E(\mathbf{k})$ is contained in the range $0 < k < \pi/a$, and only this section needs to be plotted. The point $k = \pi/a$ is called the *Brillouin zone boundary*. At these points, the wave function is a standing wave and the gradient of $E(\mathbf{k})$ vanishes. Whether the energy reaches a maximum or a minimum depends on the symmetry of the band.

We are interested in the minimum and maximum energies of the crystal bands, since these determine the band gap, and in the form of $E(\mathbf{k})$ near to these stationary points.

3.3.2.   *Conduction band*

For the conduction band, the minimum of the energy $E(\mathbf{k})$ may occur at $\mathbf{k} = 0$ or another value of $\mathbf{k}$ corresponding to an important direction in the crystal. Near to a minimum of energy at $\mathbf{k} = \mathbf{k}_{0c}$, it is often convenient to expand the $E(\mathbf{k})$ in powers of $(\mathbf{k} - \mathbf{k}_{0c})$ and approximate $E(\mathbf{k})$ by

$$E(\mathbf{k}) = E_{c0} + \frac{\hbar^2 |\mathbf{k} - \mathbf{k}_{0c}|^2}{2m_c^*} \tag{3.2}$$

where $E_{c0} = E_c(\mathbf{k}_0)$, m_c^* is a parameter with the dimensions of mass and is defined from the band structure by

$$\frac{1}{m_c^*} = \frac{1}{\hbar^2} \frac{\partial^2 E_c(k)}{\partial k^2} . \tag{3.3}$$

This is the *parabolic band approximation*. The *effective mass* m_c^* is analogous to the mass of a free electron m_0, but differs from m_0 through the different forces experienced by an electron within a crystal lattice. m_c^* can be greater or less than m_0; large values of m_c^* imply that the conduction electrons are strongly influenced by the atomic potentials. m_c^* describes how the momentum **P** conduction electron responds to an applied force, **F**

$$\mathbf{F} = m_c^* \frac{d\mathbf{p}}{dt} . \tag{3.4}$$

The value of m_c^* is determined by the atomic potentials. The effective mass is one way in which the $u_i(r)$ enter into the electron dynamics in this approximation. In general m_c^* should be a *tensor* — dependent on the direction of the applied force and the directions of the crystal axes. However, close to the band minimum the band structure is often isotropic and m_c^* can be approximated by a constant.

The two terms in Eq. 3.2 represent the potential and kinetic energy of a conduction band electron with wavevector **k**. It can be used to derive the electron velocity,

$$\mathbf{v} = \nabla_\mathbf{k} E(\mathbf{k}) = \frac{\hbar(\mathbf{k} - \mathbf{k}_{0c})}{m_c^*} \tag{3.5}$$

and momentum

$$\mathbf{p} = m_c^* \mathbf{v} = \hbar(\mathbf{k} - \mathbf{k}_{0c}) . \tag{3.6}$$

3.3.3. *Valence band*

In the valence band, we require an expression for the energy of the hole near to the valence band maximum. Since a hole represents the absence of an electron, and electrons prefer to have as little energy as possible, the most stable situation for a hole is where the electron energy is a maximum. Therefore holes are most likely to exist near to a valence band maximum, and have kinetic energy which increases as $E(\mathbf{k})$ is reduced. For a hole near a VB maximum at $\mathbf{k} = \mathbf{k}_{v0}$, the energy can be approximated by

$$E(\mathbf{k}) = E_{v0} - \frac{\hbar^2 |\mathbf{k} - \mathbf{k}_{v0}|^2}{2m_v^*} , \tag{3.7}$$

where $E_{v0} = E_v(\mathbf{k}_{v0})$, resulting in a hole velocity of

$$\mathbf{v} = -\frac{\hbar(\mathbf{k} - \mathbf{k}_{v0})}{m_v^*} \tag{3.8}$$

and momentum

$$\mathbf{p} = -\hbar(\mathbf{k} - \mathbf{k}_{v0}) \tag{3.9}$$

where the VB effective mass, m_v^* is defined through

$$\frac{1}{m_v^*} = -\frac{1}{\hbar^2}\frac{\partial^2 E_v(k)}{\partial k^2} \tag{3.10}$$

so that the mass is normally positive. In general, the effective masses of electrons and holes in the same semiconductor are different, because the curvature of the conduction and valence bands are different.

3.3.4. *Direct and indirect band gaps*

The minimum amount of energy which will promote an electron from the valence band into the conduction band is called the fundamental band gap, E_g. If the CB minimum and VB maximum occur at the same value of $\mathbf{k}$, then a photon of energy E_g is sufficient to create an electron-hole pair. This type of semiconductor is called a *direct band gap* material.

Now, if the CB minimum and VB maximum occur at *different* values of $\mathbf{k}$, a photon with energy E_g is *not* on its own sufficient to create an electron-hole pair. Promoting an electron from the VB maximum to the CB minimum would cause a change in its momentum, of $\eta(\mathbf{k}_{c0} - \mathbf{k}_{v0})$. Momentum has to be conserved in a crystal, but since photons possess virtually no momentum, that extra momentum must be supplied by something else. Usually it is supplied by a *phonon* — a lattice vibration — of the correct momentum. The phonon gives up its momentum to the electron at the moment of photon absorption so that both energy and momentum are conserved. This type of semiconductor is called an *indirect band gap* material. Clearly, photon absorption can only happen in indirect gap materials if there are enough phonons available with the required value of $\mathbf{k}$. This means that optical absorption is generally weaker for indirect than direct gap materials, and that it is more heavily dependent on temperature. The dependence of absorption on photon energy is also more gradual, as will be discussed in Chapter 4.

For device physics purposes, usually only the CB minimum and VB maximum energy are represented, in an energy–distance band diagram such as Fig. 3.7. This gives us no information on the $\mathbf{k}$ dependence of the band structure, and does not tell us whether the band gap is direct or indirect.

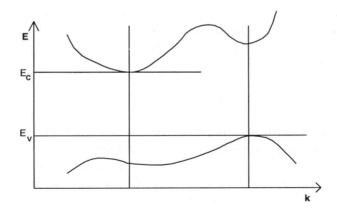

Fig. 3.6.   Band structure diagram showing indirect band gap. The maximum energy of the valence band occurs at a different wavevector to the minimum of the conduction band. Excitation of an electron across the fundamental band gap requires a change in electron momentum. This may be supplied by a phonon.

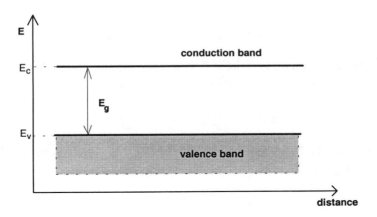

Fig. 3.7.   Energy–distance band diagram of semiconductor.

3.3.5.   *Density of states*

According to the Pauli exclusion principle, each quantum state can support only two electrons of different spin. Since each quantum state in a crystal is defined by a unique value of $\mathbf{k}$ ($\mathbf{k}$ is a 'good quantum number'), there should be two electrons per $\mathbf{k}$ value. A crystal of volume $L \times L \times L$ can support $(L/2\pi)^3$ different values of $\mathbf{k}$, and hence there are $(1/2\pi)^3$ different

k states per unit crystal volume. This follows from the requirement that $kL/2\pi$ must be an integer for a wavefunction confined in a region of width L. Including spin degeneracy, the density of electron states per unit crystal volume, $g(\mathbf{k})$, is

$$g(\mathbf{k})d^3\mathbf{k} = \frac{2}{(2\pi)^3}d^3\mathbf{k}. \qquad (3.11)$$

It is assumed that the electron levels are close enough for the integral over **k** to be a good approximation to the sum of **k** states.

It is generally much more useful to know the density of states (DOS) in terms of energy, rather than wavevector. We use the band structure $E(\mathbf{k})$ to convert the integral over wavevector space in Eq. 3.11 into an integral over energy. If the band structure is isotropic about the band minimum, assumed here to be at $\mathbf{k} = 0$, we can write

$$g(\mathbf{k})d^3\mathbf{k} = g(\mathbf{k}) \times 4\pi k^2 dk \qquad (3.12)$$

whence

$$g(E)dE = g(\mathbf{k}) \times 4\pi k^2 \times \frac{dk}{dE}dE. \qquad (3.13)$$

(If the band minimum is at $\mathbf{k}_0 \neq 0$ we simply expand about $(\mathbf{k} - \mathbf{k}_0)$. The result for $g(E)$ is the same.) Substituting for $g(\mathbf{k})$,

$$g(E) = \frac{2}{(2\pi)^3} \times 4\pi k^2 \times \frac{dk}{dE}. \qquad (3.14)$$

In the parabolic band approximation, $E(k)$ for an electron in the CB is given by Eq. 3.2. From this we obtain k^2

$$k^2 = \frac{2m_c^*}{\hbar^2}(E - E_{c0}). \qquad (3.15)$$

Differentiating Eq. 3.15 to find $\frac{dk}{dE}$ and substituting in 3.14 we find the DOS for the CB,

$$g_c(E) = \frac{1}{2\pi^2}\left(\frac{2m_c^*}{\hbar^2}\right)^{3/2}(E - E_{c0})^{3/2}. \qquad (3.16)$$

In a similar way we obtain the valence band density of states

$$g_v(E) = \frac{1}{2\pi^2}\left(\frac{2m_v^*}{\hbar^2}\right)^{3/2}(E_{v0} - E)^{3/2}. \qquad (3.17)$$

The parabolic band approximation has only limited validity for real materials. At energies far from the principal band minimum (or maximum) electron states at other symmetry points in the band structure must be included, distorting the DOS from the form in Eqs. 3.16 and 3.17. Where the parabolic band approximation is really inappropriate we should use the form of $E(\mathbf{k})$ in Eq. 3.14.

Excitons

So far we have treated the electron and hole states independently. We have not considered interactions between charge carriers which may influence the density of states. An important effect is the Coulombic interaction between electrons and holes, which gives rise to bound states called *excitons*. Excitonic states are built up from combinations of electron and hole states of the same $\mathbf{k}$. Physically they may be stationary ($\mathbf{k} = 0$) or mobile within the crystal. The exciton energies may be calculated by considering the Coulombic interaction as a perturbation of the independent particle crystal potential. A basis state of the form

$$\Psi_{\text{ex}} = \psi_{\text{e}}(\mathbf{k}, \mathbf{r})\psi_{\text{h}}(\mathbf{k}, \mathbf{r}')$$

then satisfies a hydrogenic effective mass equation

$$-\frac{\eta^2}{2\mu^*}\nabla^2 \Psi_{\text{ex}} - \frac{q^2}{4\pi\varepsilon_{\text{s}}|\mathbf{r} - \mathbf{r}'|} = E_{\text{ex}}\Psi_{\text{ex}}$$

where μ^* is the reduced effective mass of the electron-hole pair, ε_{s} the dielectric permittivity of the semiconductor and $-E_{\text{ex}}$ is the binding energy of the excitonic state. This has solutions for E_{ex}

$$E_{\text{ex}} = \frac{\mu^*}{m_0}\frac{\varepsilon_0^2}{\varepsilon_{\text{s}}^2}\frac{Ryd}{l^2} \qquad \text{where } l = 1, 2, 3, \ldots$$

where Ryd is the Rydberg energy: The electron and hole densities of states are thus modified by the presence of a series of bound states. Although excitonic states are not relevant for electrons or holes in isolation, the excitonic levels are usually drawn on band diagrams as a series of levels below the conduction band edge. The binding energies are greatly reduced compared to the hydrogen atom on account of the screening effect of the semiconductor. In a typical semiconductor, the binding energy is a few meV, so the excitons are likely to be ionised at room temperature. Excitonic states at band minima and maxima are important in the optical properties for semiconductors, particularly in low dimensional structures. In molecular materials, excitonic states are very important and dominate over band states, so that this perturbation approach is not appropriate.

Box 3.3. Low dimensional systems

The above treatment of density of states is based on a material system with isotropic symmetry and would be invalid for highly non-isotropic systems. One important such class are *low dimensional structures* where carriers are confined in one, two or three dimensions by the presence of quantum heterostructures such as quantum wells, quantum wires and quantum dots. In these cases the electron motion is strongly quantised in the confined directions, increasing the minimum energy above the minimum which applies in the bulk. Propagating $\mathbf{k}$ states can only exist in the unconfined directions, and so the sum over $\mathbf{k}$ states in Eq. 3.14 must be taken in two, one or zero directions rather than three. The result is that the form of $g(E)$ is changed. For two dimensional systems, the DOS, now per unit area, is the step function

$$g_{2D}(E) = \left(\frac{m_c^*}{\hbar^2}\right)\theta(E - E_c) \tag{3.18}$$

where $\theta(E)$ is the Heaviside function and E_c represents the new, shifted, minimum energy state. In one dimensional systems we have the DOS per unit length

$$g_{1D}(E) = \frac{1}{2\pi}\left(\frac{2m_c^*}{\hbar^2}\right)^{1/2}(E - E_c)^{-1/2}. \tag{3.19}$$

These configurations are relevant to some of the novel photovoltaic materials and device designs, which will be discussed in Chapter 10.

3.3.6. *Electron distribution function*

To calculate the densities of electrons and holes from the density of states, we need to know how the states are filled. Let's define a distribution function, $f(\mathbf{k}, \mathbf{r})$, such that f is the probability that at a point $\mathbf{r}$ the electron state of wavevector $\mathbf{k}$ is occupied. Then the density of electrons in a small volume $d^3\mathbf{k}$ around $\mathbf{k}$ is given by

$$n(\mathbf{r})d^3\mathbf{k} = g(\mathbf{k})f(\mathbf{k},\mathbf{r})d^3\mathbf{k}$$

and the total electron density in the conduction band by the sum over $\mathbf{k}$ states

$$n(\mathbf{r}) = \int_{\text{conduction band }\mathbf{k}} g_c(\mathbf{k})f(\mathbf{k},\mathbf{r})d^3\mathbf{k}. \tag{3.20}$$

Since a hole is an unoccupied electron state, the probability of a hole state being occupied is simply $1 - f(\mathbf{k},\mathbf{r})$. Thus the density of holes in the valence band is given by

$$p(\mathbf{r}) = \int_{\text{valence band }\mathbf{k}} g_v(\mathbf{k})(1 - f(\mathbf{k},\mathbf{r}))d^3\mathbf{k}. \tag{3.21}$$

3.3.7. *Electron and hole currents*

In the effective mass approximation, used so far, electron and hole densities are represented by probability distributions of plane waves. To find the electron and hole *currents*, we need to weight each probability carrier density by its group velocity. Now, by analogy with the result for free electrons, the group velocity of an electron of effective mass m_c^* in state $\mathbf{k}$ is $\frac{\eta\mathbf{k}}{m_c^*}$. The net electron current density in the conduction band should therefore be given by the sum of contributions over k states,

$$J_n(\mathbf{r}) = -\frac{q\hbar}{m_c^*}\int_{\text{conduction band }\mathbf{k}} \mathbf{k}g_c(\mathbf{k})f(\mathbf{k},\mathbf{r})d^3\mathbf{k} \tag{3.22}$$

and the hole current density in the conduction band by

$$J_p(\mathbf{r}) = \frac{q\hbar}{m_v^*}\int_{\text{valence band }\mathbf{k}} \mathbf{k}g_v(\mathbf{k})(1 - f(\mathbf{k},\mathbf{r}))d^3\mathbf{k} \tag{3.23}$$

where we make use of the fact that current is parallel to hole flow, and antiparallel to electron flow.

So, in order to calculate the densities and dynamics of electrons and holes in our semiconductor we need to know the distribution function $f(\mathbf{k},\mathbf{r})$ as well as the DOS. In general, f is position dependent and relies upon (external) factors which may vary within the material. For the special case of a system in thermal equilibrium, however, this distribution is spatially invariant and described by *Fermi Dirac statistics*. This is treated in the next section.

3.4. Semiconductor in Equilibrium

3.4.1. *Fermi Dirac statistics*

At absolute zero the electrons have no kinetic energy and always occupy the lowest available levels, filling up the available states in order of increasing energy. The energy up to which states are filled is called the *Fermi energy*, E_F. At finite temperatures the electrons have some kinetic energy, and some of them are excited into states above E_F, leaving some states below E_F unoccupied.

Thermal equilibrium means that there are no *net* exchanges of particles or energy between different points in the semiconductor or between the semiconductor and its surroundings. Every point in the semiconductor is at the ambient temperature, T, so that all carriers have the same average kinetic or *internal* energy of $\frac{3}{2}k_BT$ where k_B is Boltzmann's constant. The distribution of particles has settled so that that there are no net particle flows and E_F is the same at all points. In these equilibrium conditions, Fermi Dirac statistics describe the most stable energetic distribution of electrons which is consistent with their internal energy (k_BT) and their statistical potential energy (E_F).

For this special case, f is dependent upon wavevector only implicitly through the electron energy, and is independent of position.

$$f(\mathbf{k}, \mathbf{r}) = f_0(E(\mathbf{k}), E_F, T) \tag{3.24}$$

where $f_0(E, E_F, T)$ is the Fermi Dirac distribution function, giving the average probability that an electron state at energy E will be occupied at some temperature T,

$$f_0(E, E_F, T) = \frac{1}{e^{(E-E_F)/k_BT} + 1}. \tag{3.25}$$

3.4.2. *Electron and hole densities in equilibrium*

The electron and hole densities can be derived from Eqs. 3.21–3.22 using f_0. Provided that the density of states can be expressed in terms of energy, it is usually more convenient to do the calculations over electron states in terms of energy. Then the number density $n(E)$ of electrons with energy in the range E to E + dE in a system with density of states g(E) is given by

$$n(E)dE = g(E)f_0(E, E_F, T)dE \tag{3.26}$$

and the total density of electrons in a conduction band of minimum energy E_c by

$$n = \int_{E_c}^{\infty} g_c(E) f_0(E, E_F, T) dE \tag{3.27}$$

where the upper limit of integration has been extended to infinity.

The density of holes in a valence band of maximum energy E_v is given by

$$p = \int_{-\infty}^{E_v} g_v(E)(1 - f_0(E, E_F, T)) dE. \tag{3.28}$$

Figure 3.8 illustrates the Fermi Dirac distribution function and its effect on the energy distribution of electrons and holes in a semiconductor with roughly parabolic bands.

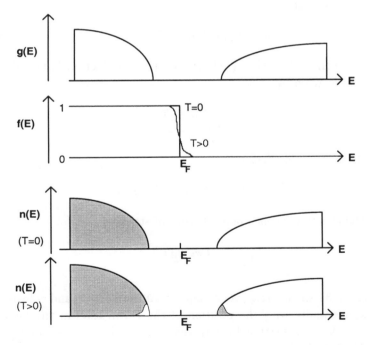

Fig. 3.8.   Density of states function $g(E)$ of a semiconductor with parabolic bands, Fermi Dirac distribution function, $f(E)$, and energy distribution, $n(E)$, of electrons (shaded area) at $T = 0$ and finite temperature.

3.4.3.   *Boltzmann approximation*

For a semiconductor at absolute zero, the valence band is completely filled and the conduction band completely empty. This means that the Fermi level must lie somewhere in the band gap. Now, if E_F is far enough from both band edges, f_0 can be well approximated by the Maxwell–Boltzmann form. In the conduction band when $E \gg E_F$,

$$f_0(E, E_F, T) \approx e^{(E_F - E)/k_B T} \tag{3.29}$$

and in the valence band where $E \ll E_F$,

$$1 - f_0(E, E_F, T) \approx e^{(E - E_F)/k_B T}. \qquad E \ll E_F \tag{3.30}$$

These approximations simplify the integrals in the expressions for n and p above (Eqs. 3.27 and 3.28). In the parabolic band approximation, using Eqs. 3.16 and 3.17 for $g(E)$, n and p can be evaluated exactly. We find for the electron density,

$$n = N_c \exp((E_F - E_c)/k_B T) \tag{3.31}$$

where the constant N_c is called the effective conduction band density of states and is given by

$$N_c = 2 \left(\frac{m_c^* k_B T}{2\pi \hbar^2} \right)^{3/2}. \tag{3.32}$$

Similarly for the hole density,

$$p = N_v \exp((E_v - E_F)/k_B T) \tag{3.33}$$

where the effective valence band density of states is

$$N_v = 2 \left(\frac{m_v^* k_B T}{2\pi \hbar^2} \right)^{3/2}. \tag{3.34}$$

Equations 3.31 and 3.33 say, very simply, that the electron and hole densities vary exponentially with the position of the Fermi level in the band gap. Moreover, they tell us that for any given material at a given temperature the product np is a constant,

$$np = N_c N_v e^{-E_g/k_B T}. \tag{3.35}$$

This relationship is used to define a constant property of the material, the *intrinsic carrier density* n_i

$$n_i^2 = np = N_c N_v e^{-E_g/k_B T} \tag{3.36}$$

n_i is equal to the density of electrons which are thermally excited into the CB of a pure semiconductor in thermal equilibrium, and is necessarily also equal to the number of holes which have been left behind.

It is sometimes convenient to write n and p in terms of n_i and an *intrinsic potential energy* E_i,

$$n = n_i \exp((E_F - E_i)/k_B T) \tag{3.37}$$

$$p = n_i \exp((E_i - E_F)/k_B T) \tag{3.38}$$

where

$$
\begin{aligned}
E_i &= \frac{1}{2}(E_c + E_v) - \frac{1}{2} kT \ln\left(\frac{N_c}{N_v}\right) \\
&= \frac{1}{2}(E_c + E_v) - \frac{3}{4} k_B T \ln\left(\frac{m_c^*}{m_v^*}\right).
\end{aligned} \tag{3.39}
$$

E_i is an energy close to the centre of the band gap, and is equal to the Fermi level of the pure semiconductor in equilibrium.

Another physically important quantity is the electron affinity of the semiconductor, χ. χ is the least amount of energy required to remove an electron from the solid. It is measured from the conduction band edge to the vacuum level E_{vac}. E_{vac} is the energy to which an electron must be raised to be free of all forces from the solid. Any spatial variations in E_{vac} are due to electrostatic fields. The conduction and valence band edges and

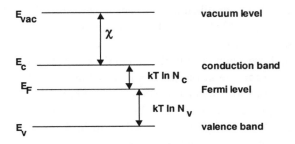

Fig. 3.9. Position of Fermi level and vacuum level on the energy band diagram.

the Fermi level can be related to E_{vac} as follows:

$$E_c = E_{vac} - \chi$$

$$E_v = E_{vac} - \chi - E_g \tag{3.40}$$

$$E_F = E_{vac} - \chi - k_B T \ln N_c = E_{vac} - \chi - E_g + k_B T \ln N_v .$$

Validity of Boltzmann approximation

The above approximations for n and p hold so long as

$$\frac{E_c - E_F}{k_B T} \gg 1 ,$$

and

$$\frac{E_F - E_v}{k_B T} \gg 1 ,$$

conditions which are usually true for a semiconductor in equilibrium. If the parabolic band approximation is not appropriate, then the form of N_c and N_v will be different, but the energy dependence of n and p will be the same as above. n and p diverge from the simple exponential form in conditions where the Fermi level approaches the lowest CB or highest VB level. This can happen *if* the semiconductor is highly doped (see below) or if there is a high density of defect states in the band gap. Then the semiconductor is said to be *degenerate*, and the full form of the Fermi Dirac distribution, Eq. 3.24, must be used to evaluate n and p.

3.4.4. Electron and hole currents in equilibrium

To find J_n and J_p in equilibrium, we need to evaluate Eqs. 3.22 and 3.23 with $f(\mathbf{k}, \mathbf{r}) = f_0(E(\mathbf{k}), E_F, T)$. Here the integral must be done over $\mathbf{k}$ states since $\mathbf{k}$ occurs explicitly in the integrand. However we can take advantage of what we know about the dependence of E on $\mathbf{k}$. Since, in the parabolic band approximation, E is an *even* function of $\mathbf{k}$, $g(E(\mathbf{k}))$ and $f_0(E(\mathbf{k}), E_F, T)$ are also even functions of k. That means that the integrand, $\mathbf{k}g(E(\mathbf{k}))f_0(E(\mathbf{k}), E_F, T)$ will be an *odd* function of k, and so the integral when taken over all $\mathbf{k}$ states will vanish. Therefore $J_n = J_p = 0$; *i.e.*, no net current flows in the semiconductor in equilibrium. That is exactly what we would expect.

From the above argument we can also see that $f(\mathbf{k}, \mathbf{r})$ must be asymmetric in $\mathbf{k}$ if the electron and hole currents are not to vanish.

3.5. Impurities and Doping

3.5.1. *Intrinsic semiconductors*

The semiconductor described so far is *intrinsic* — it is a perfect crystal containing no impurities. The only energy levels permitted are the levels of well defined $\mathbf{k}$ which arise from the overlap of the atomic orbitals into crystal bands, as described above. The properties of those bands determine the position of the Fermi level in equilibrium, E_i, and the density of carriers, n_i, which have enough thermal energy to cross the band gap and conduct electricity. They also determine conductivity since the intrinsic semiconductor has n_i electrons and n_i holes available for conduction and conductivity σ is given by

$$\sigma = q\mu_n n + q\mu_p p \qquad (3.41)$$

where μ_n, and μ_p are the mobilities of the electrons and holes, respectively, and q is the electronic charge. In equilibrium, n and p are replaced with n_i in Eq. 3.41.

At room temperature the conductivity of an intrinsic semiconductor is generally very small. For example for intrinsic silicon, $n_i = 1.02 \times 10^{10}$ cm^{-3} and $\sigma = 3 \times 10^{-6}$ Ohm^{-1} cm^{-1}, at 300 K. Conductivity increases with increasing temperature, and with decreasing band gap since n_i varies as $e^{-E_g/2k_B T}$. For example, germanium with a band gap of 0.74 eV has a conductivity of 2×10^{-2} Ohm^{-1} cm^{-1} while gallium arsenide with a band gap of 1.42 eV has a conductivity of 10^{-8} Ohm^{-1} cm^{-1}, many orders of magnitude smaller.

Now suppose that the crystal is altered by introducing an impurity atom or a structural defect. The impurity or defect introduces bonds of different strength to those which make up the perfect crystal, and therefore changes the local distribution of electronic energy levels. The altered energy levels are localised (unless the density of defects is very high), unlike the extended $\mathbf{k}$ states which make up the intrinsic conduction and valence band states.

If the impurity energy levels occur within the band gap they can affect the electronic properties of the semiconductor. Introducing occupied impurity levels above E_i increases the Fermi level, which increases the density of electrons relative to holes in equilibrium (as is clear from Eqs. 3.37 and

3.38). In the same way, unoccupied levels below E_i reduce the Fermi level
and increase the density of holes relative to electrons. The density and the
nature of the carriers in the semiconductor can thus be controlled by adding
definite amounts of impurities with energy levels close to the conduction or
valence band edge. This is called *doping*.

3.5.2. *n type doping*

A semiconductor which has been doped to increase the density of electrons relative to holes is called *n* type. (The principal charge carriers are
negative.) Occupied levels between E_i and E_c are introduced by replacing
some of the atoms in the crystal lattice with impurity atoms which possess
one too many valence electrons for the number of crystal bonds. Such impurities are called *donor* atoms because they donate an extra electron to
the lattice. An example would be an atom of phosphorus, which has five
valence electrons, in the tetravalent silicon lattice. The extra electron is not
needed in the strong directional covalent bonds which hold each atom to
its four neighbours. Therefore it is much less well bound than the other
valence electrons, and is instead held rather loosely in a Coulombic bound
state with the phosphorus atom. The donor can be ionised relatively easily,
leaving the extra electron free to move and the fixed donor atom positively
charged. For typical donors, the ionisation energy, V_n, is some meV or tens
of meV. Assuming a hydrogenic bound state, the ionisation energy is approximated by

$$V_n = \frac{m_c^* \varepsilon_0^2}{m_0 \varepsilon_s^2} Ryd$$

where ε_s is the relative dielectric constant of the material and Ryd is the
Rydberg energy, 13.6 eV. For typical crystalline semiconductors ε_s is of the
order of 10 and m_c^* less than one, meaning V_n will be reduced by a factor
of 100 compared to the Rydberg. Provided that the impurity is chosen so
that V_n is small enough, virtually all the donor atoms will be ionised at
room temperature.

In the band picture, the donor electrons have been promoted from a
level in the band gap, V_n below E_c, into the conduction band. Since the
donor states are all filled at $T = 0$, the Fermi level must now lie between
the donor level and E_c.

The density of carriers can be controlled by varying the density of
dopants, N_d. If $N_d \gg n_i$ and the donors are fully ionised at room

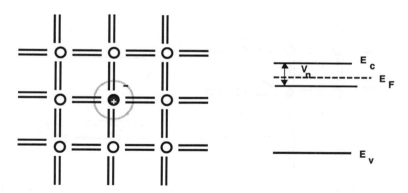

Fig. 3.10. Schematic illustration of an n type impurity in the bond (left) and band (right) picture. The extra electron is loosely bound to the donor atom. When the ionisation energy of V_n is supplied, the extra electron is freed from the donor, and the donor atom becomes positively charged. In the band picture, the electron is raised through V_n into the conduction band.

temperature, then

$$n \approx N_d \tag{3.42}$$

and

$$p = n_i^2/N_d, \tag{3.43}$$

since $np = n_i^2$ at equilibrium. The electron density is now greatly increased over its equilibrium value, while the hole density is greatly reduced. The electrons are here called the *majority* carriers and the holes the *minority* carriers. Relative to the intrinsic case, the total density of carriers is increased, and so therefore is the conductivity. Conduction in n type semiconductors is mainly by electrons.

3.5.3. *p type doping*

A semiconductor which is doped to increase the density of positive charge carriers relative to negative is called p type. It is produced by replacing some of the atoms in the crystal with *acceptor* impurity atoms, which contribute too few valence electrons for the bonds they need to form. An example would be the trivalent element boron in silicon. The acceptor becomes ionised by removing a valence electron from another bond to complete the bonding between it and its four neighbours. This releases a hole into the valence

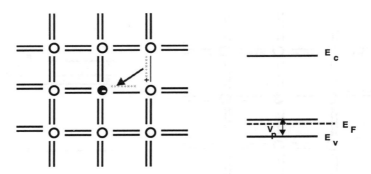

Fig. 3.11.   Schematic illustration of a p type impurity in the bond (left) and band (right) picture. The acceptor atom has one too few electrons, and becomes ionised by grabbing another electron from a neighbouring bond. The grabbed electron is more tightly bound to the impurity than to a normal crystal atom. The hole created is relatively free to travel from site to site. In the band picture, the hole is lowered through V_p into the valence band.

band. The energy of this ionised state is higher, by a small amount of energy V_p, than the highest energy of a valence band electron, and so the acceptor energy level appears in the band gap close to the valence band edge. The Fermi level now lies between the acceptor level and E_v.

These p type semiconductors contain an excess of positive carriers, holes. For a doping density of $N_a \gg n_i$ ionisable acceptors, we find

$$p \approx N_a \qquad (3.44)$$

and

$$n = n_i^2/N_a . \qquad (3.45)$$

Again the total carrier density and hence the conductivity is greatly increased, but now conduction is mainly by holes, which are the majority carriers in this case.

Notice from Eqs. 3.37, 3.38 and 3.42–3.45 how the changes in n and p are consistent with the changes in the Fermi level. In an intrinsic (*i.e.*, pure) semiconductor E_F lies roughly in the middle of the band gap. The effect of doping is to shift the Fermi level away from the centre of the band gap, towards E_c in n type material, and towards E_v in p type material.

Note also that, unlike heating and illumination, doping is a way of increasing the conductivity of the semiconductor at equilibrium without requiring a constant input of energy.

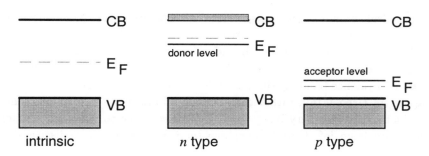

Fig. 3.12. Position of Fermi level and impurity levels relative to the conduction and valence band edges. Shaded regions indicate levels filled by electrons.

3.5.4. *Effects of heavy doping*

Heavy doping causes imperfections in the crystal structure which appear as states in the band gap close to the CB and VB edges. These states have the effect of adding a tail to the CB and VB density of states functions, and so effectively reducing the band gap. The intrinsic carrier density will increase, because thermal excitation across the band gap is easier, and this has consequences for solar cell performance. Heavy doping is also likely to increase the density of defect states which act as centres for carrier recombination or trapping.

3.5.5. *Imperfect and amorphous crystals*

As well as extrinsic impurities, semiconductors may contain intrinsic defects which introduce levels into the band gap and modify the electronic properties. Even in a perfect monocrystalline semiconductor, configurational defects arise from broken bonds and structural rearrangement at surfaces. In a polycrystalline semiconductor, made up of grains of the same crystal at various orientations, defects occur at grain boundaries. These defects may cause an otherwise intrinsic material to become slightly n or p type, by self doping. They can also be responsible for charged boundary layers near to surface or grain boundaries in deliberately doped materials, which impede carrier transport.

There is another important group of semiconductor materials, amorphous semiconductors. These have a similar structure to the crystal on the short range, but have no long range order. For instance, in amorphous silicon, while each silicon atom is bonded to four neighbours with approximate tetrahedral symmetry (just as in crystalline silicon) the arrangement of silicon atoms four or five bond lengths away appears almost random. These small deviations in bond length and orientation give rise to a spread in energy levels so that the density of states near the conduction band minimum and valence band maximum tends to decay with energy into the band gap, in an Urbach tail, rather than cutting off abruptly. Amorphous semiconductors are also very likely to contain defects due to voids and broken bonds. Intrinsic defects and amorphous semiconductors will be discussed in Chapter 8.

A very high density of defects in the band gap is undesirable, as these tend to act as centres for charge recombination (see Chapter 4). It also gives rise to an effect called Fermi level pinning, where the density of intra band gap states is too high for all the states to be ionised at room temperature, and so the Fermi level is fixed amongst them. This makes it difficult to dope the semiconductor in the usual way. Fermi level pinning can be a problem at interfaces and is mentioned in Chapter 5.

3.6.   Semiconductor under Bias

So far we have considered the semiconductor at equilibrium. One feature of the material at equilibrium is that no current flows. To understand the operation of a solar cell, we need to understand what happens to the semiconductor when it is displaced from equilibrium by exposure to light. Since current flows in the operating solar cell, by definition the device cannot be at equilibrium.

3.6.1.   *Quasi thermal equilibrium*

Suppose the semiconductor is disturbed from equilibrium by some influence which changes the population of electrons and holes. This could be exposure to light of energy $E > E_g$ which increases the density of both electrons and holes above their equilibrium values, or it could be the electrical injection of electrons and holes through an applied electric bias. In either case, n and p have been disturbed so that they are no longer described by the Fermi Dirac equilibrium distribution function and $np = n_i^2$ is no longer true.

The electron distribution is now governed by the *general* distribution function $f(\mathbf{k}, \mathbf{r})$ introduced in Eq. 3.20 above. Moreover, different distribution functions, $f_c(\mathbf{k}, \mathbf{r})$ and $f_v(\mathbf{k}, \mathbf{r})$, should apply for electrons in conduction and valence bands. f_c and f_v should be position dependent, because all points in the semiconductor are no longer in equilibrium, and also $\mathbf{k}$ dependent because the applied bias may be directional and may favour occupation of some $\mathbf{k}$ states rather than others of the same E.

Now — and here comes one of the great simplifications of semiconductor physics — if the disturbance is not too great, or not changing too quickly, the populations of electrons and holes each relax to achieve a state of *quasi thermal equilibrium*. What that means is that the population of electrons within the conduction band distribute themselves as though they were at equilibrium, with a common Fermi level and temperature, and the holes within the valence band arrange themselves as though they shared another, different Fermi level and possibly a different temperature. In quasi thermal equilibrium,

$$f_c(\mathbf{k}, \mathbf{r}) \approx f_0(E, E_{F_n}, T_n)$$

and

$$f_v(\mathbf{k}, \mathbf{r}) \approx f_0(E, E_{Fp}, T_p) \, .$$

The new apparent Fermi levels for electrons and holes are called the electron and hole *quasi Fermi levels*, E_{F_n} and E_{Fp}, respectively.

The approximation is possible because relaxation *within* each band is so much faster than relaxation *between* the bands. (Carriers relax within the bands mainly by scattering from the lattice, with the emission and absorption of phonons. This occurs on a time scale of 10^{-12}–10^{-15} s. Between bands, the carriers relax only by interacting with another carrier, with the emission of a photon or with a deep trap. Relaxation over these energies is much less likely because of the scarcity of high energy phonons and photons under normal conditions, and time scales are longer, typically 10^{-6}–10^{-9} s.) It is possible for a disturbance — such as optical generation — to be fast on the timescale of relaxation between bands, so that it continuously disturbs the system from equilibrium, but slow on the timescale of relaxation within bands.

Considering only the conduction band for the moment, $f_c(k, r)$ cannot be *completely* described by the quasi thermal equilibrium distribution however; since, as we have seen above, f_0 is symmetric in $\mathbf{k}$, implying that no current flows. To deal with this, a small additional term, f_A, which is *antisymmetric* in $\mathbf{k}$, is added to f_0. The approximation is written,

$$f_c(\mathbf{k}, \mathbf{r}) = f_0(E, E_{F_n}, T_n) + f_A(\mathbf{k}, \mathbf{r}) \qquad (3.46)$$

where

$$f_0(E, E_{F_n}, T_n) = \frac{1}{e^{(E-E_{F_n})/k_B T_n} + 1}.$$ (3.47)

f_A can be found by setting up and solving a book-keeping equation for f called the Boltzmann Transport equation (see Box 3.4). Note that because of the well defined parity of f_0 and f_A, only f_0 contributes to the expressions for n and p, while only f_A contributes to J_n and J_p.

3.6.2. *Electron and hole densities under bias*

Assuming quasi thermal equilibrium, the electron and hole densities in a semiconductor under bias are given, using Eqs. 3.37 and 3.38, by

$$n = n_i e^{(E_{F_n} - E_i)/k_B T_n}$$ (3.48)

and

$$p = n_i e^{(E_i - E_{F_p})/k_B T_p}$$ (3.49)

or, in terms of the conduction and valence band energies using Eqs. 3.31 and 3.34, by

$$n = N_c e^{-(E_c - E_{F_n})/k_B T_n}$$ (3.50)

and

$$p = N_v e^{-(E_{F_p} - E_v)/k_B T_p},$$ (3.51)

where T_n and T_p are the electron and hole effective temperatures. In principle, T_n and T_p may be different from the ambient temperature T. Electrons with $T_n > T$ are called 'hot' electrons and can arise in situations where the carriers gain excess kinetic energy from strong electric fields. However, the conditions giving rise to hot carriers do not usually apply in photovoltaic devices, and henceforth we will assume that $T_n = T_p = T$. (We will come back to hot electrons in Chapter 10.) It is assumed in Eqs. 3.48–3.51 that $E_c - E_{F_n} \gg k_B T$ and $E_{F_p} - E_v \gg k_B T$, so that Boltzmann statistics apply.

Equations 3.48 and 3.49 are the same as Eqs. 3.37 and 3.38 except that $E_{F_n} \neq E_{F_p} \neq E_F$. We say that the Fermi levels are *split* (see Fig. 3.13). The difference in quasi Fermi levels,

$$\Delta\mu = E_{F_n} - E_{F_p}$$ (3.52)

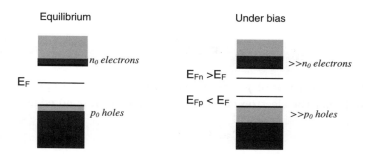

Fig. 3.13. Fermi levels at and away from equilibrium.

is the difference in chemical potentials referred to in Chapter 2, and the size of the splitting, $\Delta\mu$, depends upon the intensity of the disturbance. The electron hole product is now given by

$$np = n_i^2 e^{\Delta\mu/k_BT} \,. \tag{3.53}$$

In general, E_{F_n} and E_{Fp} are functions of position. At any point within the semiconductor we will suppose there is a *local* quasi thermal equilibrium and we can define *local* quasi Fermi levels.

3.6.3. *Current densities under bias*

To find J_n and J_p we need to find an approximation for the antisymmetric part of f. This is done by considering the different influences on f.

Box 3.4. Boltzmann Transport equation and the relaxation time approximation

In this box we derive a form for f_A based on a consideration of the processes by which carriers are removed from any particular $\mathbf{k}$ state.

Since f_c is a function of $\mathbf{k}$, $\mathbf{r}$ and t, the rate of change of f is given by

$$\frac{df_c}{dt} = \frac{d\mathbf{r}}{dt} \cdot \nabla_r f_c + \frac{d\mathbf{k}}{dt} \cdot \nabla_k f_c + \frac{\partial f_c}{\partial t} \,. \tag{3.54}$$

The terms in this equation represent the various ways in which f_c can change. In the following we will use the definitions of velocity

$$\mathbf{v} = \frac{d\mathbf{r}}{dt} \,, \tag{3.55}$$

and force

$$\mathbf{F} = \hbar \frac{d\mathbf{k}}{dt}.$$

$$(3.56)$$

To find the derivatives of f with respect to $\mathbf{r}$ and $\mathbf{k}$ we will approximate f_c by f_0 and refer the electron energy to the conduction band edge,

$$E = E_c + E(\mathbf{k}, \mathbf{r}).$$

$$(3.57)$$

(Here we have assumed, for clarity, that the conduction band minimum is at $\mathbf{k}_0 = 0$. The result is the same for $\mathbf{k}_0 \neq 0$.) Then

$$\nabla_{\mathbf{r}} f_c \approx -\frac{f_0}{kT} \nabla_{\mathbf{r}}(E_c - E_{F_n})$$

$$(3.58)$$

and

$$\nabla_{\mathbf{k}} f_c \approx -\frac{f_0}{kT} \nabla_{\mathbf{k}} E = -\frac{\hbar f_0}{k_B T} \mathbf{v}$$

$$(3.59)$$

where we have used the definition of group velocity. (We have also assumed that there is no gradient in temperature, as discussed above. In the most general case $\nabla T \neq 0$ and the resulting expression for current contains a term driven by temperature gradients.) Substituting back into Eq. 3.54 we obtain

$$\frac{df_c}{dt} = -\frac{f_0}{k_B T}(\mathbf{v} \cdot \nabla_{\mathbf{r}}(E_c - E_{F_n}) + \mathbf{F} \cdot \mathbf{v}) + \frac{\partial f_c}{\partial t}$$

$$= \frac{f_0}{k_B T} \mathbf{v} \cdot \nabla_{\mathbf{r}} E_{F_n} + \frac{\partial f_c}{\partial t}.$$

$$(3.60)$$

where we have related $\nabla_{\mathbf{r}} E_c$ to the force on the electron to cancel out the first and third terms.

To solve this for f_c we need to make two assumptions about the processes which change f_c. One, relaxation events within a band within a band are much more frequent than exchanges of carriers between bands. As mentioned above, relaxation events within a band are usually collisions with the lattice and are fast because of the abundance of phonons. Therefore

$$\frac{\partial f_c}{\partial t} \approx \left. \frac{\partial f_c}{\partial t} \right|_{\text{collisions}}.$$

Two, the distribution relaxes exponentially towards the quasi equilibrium function with a characteristic time constant τ so that

$$\left.\frac{\partial f_c}{\partial t}\right|_{\text{collisions}} = -\frac{(f_c - f_0)}{\tau} \tag{3.61}$$

This is the *relaxation time approximation*. Then we can solve Eq. 3.60 in the steady state to obtain

$$f_c = f_0 \left(1 - \frac{\tau \mathbf{v}}{k_B T} \cdot \nabla_{\mathbf{r}} E_{F_n}\right) \tag{3.62}$$

whence

$$f_A = -f_0 \frac{\tau \mathbf{v}}{k_B T} \cdot \nabla_{\mathbf{r}} E_{F_n} . \tag{3.63}$$

Substituting for f_c into Eq. 3.22 for J_n we find

$$J_n(\mathbf{r}) = \left\{ \frac{q}{k_B T} \int_{\text{conduction band } \mathbf{k}} \tau \left(\frac{\hbar \mathbf{k}}{m_c^*}\right)^2 g_c(\mathbf{k}) f_0(E(\mathbf{k})) d^3\mathbf{k} \right\}$$
$$\times \nabla_{\mathbf{r}} E_{F_n} . \tag{3.64}$$

Finally, replacing the term in brackets by $\mu_n n$, where μ_n is the electron *mobility*, and n defined from f_0 as above, (Eq. 3.21) we have

$$\boxed{J_n(\mathbf{r}) = \mu_n n \nabla_{\mathbf{r}} E_{F_n}} . \tag{3.65}$$

So, the electron current is proportional to the gradient in the electron quasi Fermi level.

A similar treatment for the valence band distribution function f_v produces the hole current density

$$\boxed{J_p(\mathbf{r}) = \mu_p p \nabla_{\mathbf{r}} E_{F_n}} . \tag{3.66}$$

Henceforth we will use ∇ to mean $\nabla_{\mathbf{r}}$.

Using the Boltzmann Transport equation to solve for the non-equilibrium distribution function (see Box 3.4), we obtain the following relations for the electron and hole current densities in a semiconductor under bias:

$$J_n(\mathbf{r}) = \mu_n n \nabla E_{F_n} \tag{Eq. 3.65}$$

$$J_p(\mathbf{r}) = \mu_p p \nabla E_{F_p} , \tag{Eq. 3.66}$$

that is, the current density at any point is proportional to the relevant carrier density and the gradient of the relevant quasi Fermi level. We say the current is driven by the gradient in the quasi Fermi level. In a more general case, Eqs. 3.65 and 3.66 would contain terms in ∇T_n and ∇T_p and the currents would be driven also by temperature gradients.

At any point $\mathbf{r}$ the net current is given by the sum of the electron and hole currents at that point:

$$J(\mathbf{r}) = J_n(\mathbf{r}) + J_p(\mathbf{r}), \tag{3.67}$$

3.7. Drift and Diffusion

3.7.1. *Current equations in terms of drift and diffusion*

Equations 3.65 and 3.66 are the most compact forms for J_n and J_p. However, the physical interpretation of a quasi Fermi level gradient may be somewhat clearer if we use the results, valid within the Boltzmann approximation, that

$$\nabla E_{F_n} = (\nabla E_c - kT\nabla \ln N_c) + \frac{k_B T}{n}\nabla n \tag{3.68}$$

and

$$\nabla E_{F_p} = (\nabla E_v + kT\nabla \ln N_v) - \frac{k_B T}{p}\nabla p. \tag{3.69}$$

(These can be obtained by differentiating Eqs. 3.31 and 3.33.) The gradient in the conduction or valence band edge is provided by the electrostatic field F and the gradients in the electron affinity χ and band gap, E_g

$$\nabla E_c = qF - \nabla\chi \tag{3.70}$$

and

$$\nabla E_v = qF - \nabla\chi - \nabla E_g \tag{3.71}$$

where we have used Eq. 3.40 and the definition of electrostatic field

$$F = \frac{1}{q}\nabla E_{vac}. \tag{3.72}$$

Substituting for ∇E_{F_n} and ∇E_{F_n} into Eqs. 3.65 and 3.66 we have

$$J_n(\mathbf{r}) = qD_n\nabla n + \mu_n n(qF - \nabla\chi - k_B T\nabla \ln N_c). \tag{3.73}$$

and

$$J_p(\mathbf{r}) = -qD_p\nabla p + \mu_p p(qF - \nabla\chi - \nabla E_g + k_BT\nabla\ln N_v). \quad (3.74)$$

Thus the gradients in the electron affinity, band gap and effective band densities of states provide an additional *effective* electric field to the electrostatic field F.

In compositionally invariant material, only the electrostatic field F is present and so

$$J_n(\mathbf{r}) = qD_n\nabla n + q\mu_n Fn \quad (3.75)$$

and

$$J_p(\mathbf{r}) = -qD_n\nabla p + q\mu_p Fp. \quad (3.76)$$

Above we have used the 'Einstein' relations which relate mobility to the diffusion constant for either carrier, and are valid under low field conditions

$$\mu_n = \frac{qD_n}{k_BT}, \qquad \mu_p = \frac{qD_p}{k_BT}. \quad (3.77)$$

Now, J_n and J_p each resolve into two contributions: a *drift* current (sometimes called migration) where carriers are driven by an electric field, and a *diffusion* current where they are driven by a concentration gradient. In the first case the carriers move so as to minimise their electrostatic potential energy, and in the second to minimise their statistical potential energy.

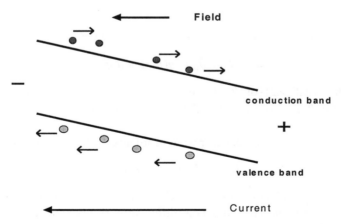

Fig. 3.14. Drift. Carriers flow under an *electric field* in order to reduce their electrical *potential* energy. Notice that electrons and holes drift in opposite directions, both producing currents in the direction of the field.

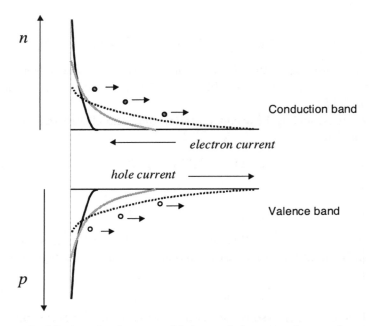

Fig. 3.15. In diffusion, the electron and hole populations spread out under a concentration gradient to reduce their statistical potential energy. The grey and dotted lines indicate how the electron and hole density may develop after generation close to the left hand surface. Note that the currents produced by electron and hole diffusion have opposite signs. Unless the electron and hole mobilities are different, the net current may be zero.

In the presence of a drift field only, the total current is

$$J = J_\mathrm{n} + J_\mathrm{p} = q(\mu_\mathrm{n} n + \mu_\mathrm{p} p)F \tag{3.78}$$

where the coefficient of F is the conductivity of the semiconductor, as given in Eq. 3.41. Note that the field drives electrons and holes in opposite directions.

In the presence of a concentration gradients only we have,

$$J_\mathrm{n} = q(D_\mathrm{n} \nabla n - D_\mathrm{p} \nabla p) \tag{3.79}$$

If the electron and hole gradients are similar, as may be the case under illumination, the electron and hole diffusion currents tend to cancel each other out. A *net* diffusion current usually arises only when the electron and hole carrier gradients are very different. This can be achieved in certain device configurations, such as a *p–n* junction. In that case, currents are dominated by minority carrier diffusion. This will be discussed in detail in Chapter 6.

3.7.2. *Validity of the drift-diffusion equations*

The drift-diffusion current equations in the form 3.75 and 3.76 are very commonly used for the electron and hole currents in device physics. However, they depend upon several assumptions. It is worth reviewing these here, and considering some exceptions.

Assumptions used:

- Electron and hole populations each form a quasi thermal equilibrium with a characteristic Fermi level and temperature

This is one of the fundamental tenets of semiconductor physics. It would not be valid if for instance, the bands were too narrow, too close together, or if there was too high a density of intra band gap states. In those conditions we might not really have a semiconductor!

- Electron and hole temperatures are always the same as the lattice temperature

This fails in conditions where electrons or holes gain extra kinetic energy; for example, through acceleration through large electric fields. Then the hot electrons and holes need to be described by spatially varying temperatures T_n and T_p, and a term proportional to the temperature gradient appears in the electron and hole currents. Such high fields are not usually encountered in photovoltaic devices. However, it is theoretically possible that hot carrier effects may arise through quantum confinement in low dimensional structures. This will be raised in Chapter 10.

- Changes in state occupancy are much more likely to be due to scattering collisions within a band than to generation, recombination or trapping events which remove a carrier from that band. This is the relaxation time approximation.

This may fail in materials containing a high density of defects, where trapping, detrapping and recombination events are common. Then the relaxation time approximation is not valid. Another special case arises at an abrupt boundary between two different semiconductors (a heterojunction).

- Electron and hole states can be described by a quantum number **k**

This is used in the definition of the electron and hole currents above. It fails for non-crystalline materials, with the consequence that carrier mobilities cannot be rigorously defined from the expectation value of the scattering time. Nevertheless, analogous expressions for electron and hole currents can be used in amorphous and defective materials, but with empirical expressions for the mobilities.

- Boltzmann approximation ($E_c - E_{F_n} \gg k_B T$; $E_{F_p} - E_v \gg k_B T$)

This was made in obtaining the drift diffusion forms from the quasi Fermi level gradients. In degenerate semiconductors (where the Boltzmann approximation fails) the definitions in terms of Fermi level gradients are still valid and should be used.

- Compositional invariance

This was made in replacing the band edge gradient with F. For a compositionally varying material, contributions to the current will also arise from any variation in the electron affinity, band gap, and effective mass, and can be included by using Eqs. 3.73 and 3.74.

3.7.3. *Current equations for non-crystalline solids*

The drift diffusion picture is clearly not valid in the case of defective crystalline or amorphous materials, when the density of states in the band gap is high. Then the following effects may apply:

- the number density of electrons and holes may be differently sensitive to temperature, electric field and illumination;
- the mobility may be temperature, field and carrier density dependent;
- there may be a contribution to the current which comes from conduction between localised states in the band gap.

There are a number of different ways of describing the currents in defective materials. A thorough description is beyond the scope of this chapter. It is often assumed that the currents due to electrons and holes moving in delocalised states in the conduction and valence bands can be described by the Boltzmann transport theory, *i.e.*, drift diffusion equations apply. There may be an additional current due to electron (or hole) motion between localised states in the band gap,

$$J = J_n + J_p + \sum_{\text{localised states } i} J_i. \tag{3.80}$$

This localised state current can be described with the aid of an energy dependent mobility, $\mu(E)$, in analogy to the definition of current in a crystal above

$$\sum_i J_i = \sum_i \mu(E_i) f(E_i) g(E_i) \tag{3.81}$$

where $g(E_i)$ is the density of localised states and $f(E)$ is the (usually Fermi–Dirac) distribution function. The form of $\mu(E)$ depends on the mechanism of transfer between localised states. For example, for thermally assisted hopping it should be thermally activated.

Delocalised conduction or valence band states can be distinguished from localised, intra band gap states by a mobility edge. In practice, this is the energy at which the mobility becomes temperature activated. These will be discussed in more detail in Chapter 8.

3.8. Summary

Semiconductors are suitable materials for photovoltaics on account of the band gap between occupied and unoccupied bands which can be bridged by a visible photon, and the ease of charge transport through the crystal bands. In crystalline semiconductors, electrons promoted to the conduction band can be treated like nearly free particles with a well-defined wavevector and an effective mass which reflects the effect of the lattice atoms. Absent electrons in the valence band are known as holes and can be treated as positively charged particles with an effective mass determined by the valence band structure. Electrons and holes can be treated approximately as independent particles. The band structure is the relationship between the energy of the electron and its wavevector. The most important points are the valence band maximum and conduction band minimum. The form of the band structure is determined by the crystal structure but it is approximately parabolic near to these critical points.

In equilibrium, states are filled by electrons according to the Fermi Dirac distribution function. In a pure semiconductor the Fermi level lies within the band gap and the electron and hole densities depend exponentially upon E_F. Doping with impurities of different valence increases the density of electrons or holes in the semiconductor at equilibrium and moves E_F towards the bands. Heavy doping fixes the density of the dominant carrier. At equilibrium, electron and holes currents are always zero.

The semiconductor can be disturbed from equilibrium by the action of light or an applied electric bias. Away from equilibrium the distribution function is changed but can be approximated by a Fermi Dirac function with separate, spatially varying Fermi levels, called quasi Fermi levels, for the electron and hole. In general, different electron and hole temperatures may apply but in practical conditions the ambient temperature is usually valid.

Application of bias produces electron and hole currents. In conditions where scattering events within bands being faster than transitions between bands (the relaxation time approximation), the currents are proportional to the gradients in the respective quasi Fermi levels and to the respective mobilities. The quasi Fermi level gradient may be considered as a driving force for conduction. By resolving the quasi Fermi level gradient into electrostatic potential and carrier density gradients, the current can be expressed as a sum of a field driven 'drift' current and a statistically driven 'diffusion' current. In non-crystalline materials the same concepts of quasi Fermi level and mobility can be applied, but the effective mobility may be energy dependent.

Chapter 4

Generation and Recombination

4.1. Introduction: Semiconductor Transport Equations

The essential function of a solar cell is the generation of photocurrent. The output is determined by a balance between light absorption, current generation and charge recombination. Currents due to electrons and holes in semiconductors were discussed in Chapter 3, and Chapter 5 will deal with the mechanisms of charge separation which drive the photocurrent. This chapter is concerned with the processes of charge carrier generation and recombination. We introduce the theoretical formalism for electronic transitions in semiconductors and use this to derive the rates of photogeneration and the principal recombination mechanisms. We will derive some of the commonly used formulae for recombination rates, and attempt to show how the rates are dependent upon the electronic structure, materials properties and operating conditions. The detailed derivations will not be relevant to all readers, but are included to suggest how the formalisms can be adapted to deal with other, novel photovoltaic materials and structures.

The basic equations of device physics, the semiconductor transport equations, are based on two simple principles: that the number of carriers of each type must be conserved; and that the electrostatic potential due to the carriers charges obeys Poisson's equation. For a semiconductor containing electrons and holes, conservation of electron number requires that

$$\frac{\partial n}{\partial t} = \frac{1}{q}\nabla.\mathbf{J}_n + G_n - U_n \qquad (4.1)$$

for electrons and

$$\frac{\partial p}{\partial t} = -\frac{1}{q}\nabla.\mathbf{J}_p + G_p - U_p \qquad (4.2)$$

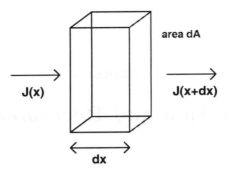

Fig. 4.1. Illustration of the continuity equations. On passing through a small volume, the charge changes by $(J(x+dx) - J(x)) \times dA$. This has to be supplied by the difference between the carrier generation and recombination rates, $(G - U) \times dx \times dA$.

for holes, where $G_{n/p}$ is the volume rate of generation of electrons (holes) and $U_{n/p}$ is the volume rate of recombination. Poisson's equation in the differential form is

$$\nabla^2 \phi = \frac{q}{\varepsilon_s}(-\rho_{\text{fixed}} + n - p) \qquad (4.3)$$

where ϕ is the electrostatic potential, ε_s is the dielectric permittivity of the semiconductor and ρ_{fixed} is the local density of fixed charge.

The continuity Eqs. 4.1 and 4.2 can be understood by considering a small volume $\delta\mathscr{V} = \delta A \times \delta x$ within the semiconductor (Fig. 4.1).

In unit time, $\delta\mathscr{V} \times G_n$ electrons are generated within the volume and $\delta\mathscr{V} \times U_n$ electrons are removed. Electrons may also be introduced or removed by currents flowing into and out of the volume. Consider only the x component of the current. In unit time $\frac{1}{q}J_n(x+\delta x) \times \delta A$ electrons leave the volume through the right hand boundary while $\frac{1}{q}J_n(x) \times \delta A$ enter from the left, causing a net change of $\frac{1}{q}\frac{\delta J_n}{\delta x} \times \delta v$ in the number of electrons. Adding the three contributions we have for the rate of change of electron density per unit volume,

$$\frac{\partial n}{\partial t} = \frac{1}{q}\frac{\partial J_n}{\partial x} + G_n - U_n .$$

In three dimensions this generalises quite readily to the form given in (4.1) above.

Equations (4.1) and (4.2) are completely general. Once we know how the J, G and U depend on n, p and ϕ, and other material or environment parameters, we have a set a set of coupled differential equations which can be solved for the three unknowns. The particular form of J, G and U depends on the material and environment. In Chapter 3 we looked at how J and n are defined for a crystalline semiconductor. In this chapter we will look at the mechanisms which determine U and G.

4.2. Generation and Recombination

Generation is an electronic excitation event which increases the number of free carriers available to carry charge. Recombination is an electronic relaxation event which reduces the number of free carriers. Generation requires an input of energy while recombination releases energy. The energy input can be provided by the vibrational energy of the lattice (phonons), light (photons) or the kinetic energy of another carrier. The released energy is taken up by these same mechanisms. For every generation process there is an equivalent recombination process. This is due to microscopic reversibility, and is an important principle in understanding the function of photovoltaic devices.

Generation may be the promotion of an electron from valence to conduction band, which creates an electron-hole pair, or from valence band into a localised state in the band gap, which generates only a hole, or from a localised state into the conduction band, which generates only an electron. For the solar cell, the most important form of generation is optical, *i.e.*, by the absorption of a photon.

Recombination is the loss of an electron or hole through the decay of an electron to a lower energy state. Again this may be from band to band, destroying an electron-hole pair, or it may be from conduction band to trap state or from trap state to valence band, removing only an electron or a hole, respectively. The energy released can be given up as a photon (*radiative* recombination), as heat through phonon emission (*non-radiative* recombination) or as kinetic energy to another free carrier (*Auger* recombination).

Thermal generation and recombination

At absolute zero in the absence of any external bias, electrons occupy all of the energy levels available up to the Fermi level, and no recombination or generation processes occur. As the temperature is raised, the lattice gains vibrational kinetic energy and some of this may be given up to an electron to promote it to a higher energy level. The promotion of an electron across the band gap is is called *thermal generation*. At the same time, electrons in excited states can relax down to vacant lower energy states and give up the energy difference as vibrational energy to the lattice. The loss of a mobile carrier in this way is *thermal recombination*. Like any other generation-recombination processes, these thermal processes can involve band to band transitions and localised state to band transitions. At finite temperatures these processes occur continually at a rate which increases with increasing temperature. In thermal equilibrium, the rate of every thermal generation process is matched exactly by the rate of the equivalent thermal recombination events. Thus the thermal generation $G_{n/p}^{th}$ is balanced by the equilibrium recombination rate $U_{n/p}^{th}$,

$$G_n^{th} = U_n^{th}$$

and

$$G_p^{th} = U_p^{th}.$$

Since we are interested in the *disturbance* of the populations from equilibrium, we need consider only the excess recombination and generation rates in Eq. (4.1), *i.e.*,

$$U_n = U_n^{total} - U_n^{th} \tag{4.4}$$

and

$$G_n = G_n^{total} - G_n^{th}.$$

Thus thermal generation is not included explicitly as a contribution to G.

For band-to-band generation and recombination processes, the rates are equal for electrons and holes and

$$U_n = U_p = U \tag{4.5}$$

and

$$G_n = G_p = G.$$

In this chapter we will first consider how electronic transition rates may be treated microscopically using Fermi's Golden Rule, in Sec. 4.3. In Sec. 4.4 we will address the process of photogeneration, and show how the absorption coefficient of a semiconductor is derived. In Sec. 4.5 we consider the different types of recombination: radiative, Auger, nonradiative via traps, and surface recombination, and show how the recombination rates are related to the material properties. Section 4.6 addresses the semiconductor transport equations again and indicates some simplified cases.

4.3. Quantum Mechanical Description of Transition Rates

4.3.1. *Fermi's Golden Rule*

In many cases electronic transitions can be described by Fermi's Golden Rule. This is an approximation, based on first order perturbation theory, of the full quantum mechanical transition rate. According to Fermi's Golden Rule, the transition *probability* per unit time from an initial filled state $|i\rangle$ of energy E_i to a final empty state $|f\rangle$ of energy E_f which differs in energy by E under the action of some perturbing Hamiltonian $\mathbf{H}'$ is given by

$$\frac{2\pi}{\hbar} |\langle i|\mathbf{H}'|f\rangle|^2 \delta(E_f - E_i \mp E). \tag{4.6}$$

The term in brackets $|\langle\rangle|^2$ is a *matrix element* coupling initial and final states. It describes the size of the interaction between the two states under the given perturbation. $\mathbf{H}'$ may represent the effect of a light field, of phonon interactions, or carrier–carrier interactions. The delta function ensures conservation of energy. A negative sign applies for excitation events, such as absorption where $E_f > E_i$, and a quantum of energy E is removed from the perturbing field, and a positive sign for relaxation events, such as recombination, where $E_f < E_i$ and a quantum E is emitted.

Box 4.1. Derivation of Fermi's Golden Rule

To obtain the transition *rate* we need to multiply by the probabilities that the initial state is occupied, $f_i = f(E_i(\mathbf{k}_i))$ and that the final state is available, $1 - f_f = (1 - f(E_f(\mathbf{k}_f)))$, where $f(E)$ is the electronic occupation function. Then the rate of transitions $i \to f$ is given by

$$r_{i \to f} = \frac{2\pi}{\hbar} |\mathbf{H}_{if}|^2 \delta(E_f - E_i - E) f_i (1 - f_f) \tag{4.7}$$

where $\mathbf{H}_{if}$ is shorthand for the matrix element. (In Eq. 4.7 an electron promotion event ($E_f > E_i$, is assumed. The analysis applies equally to the case $E_f < E_i$, in which case E takes the opposite sign.)

Provided that the transition $i \to f$ is energy conserving, then the same perturbing field can induce the *reverse* transition $f \to i$. (Stimulated emission in a solid is an example of this.) As soon as state f is partially occupied and state i is partially empty, there is a finite probability that this reverse transition will occur. By the symmetry of quantum mechanics the transition $f \to i$ is governed by the *same* matrix element H_{if}, and it has rate

$$r_{f \to i} = \frac{2\pi}{\hbar} |\mathbf{H}_{if}|^2 \delta(E_f - E_i - E) f_f (1 - f_i) \tag{4.8}$$

The *net* rate of transitions $i \to f$ is therefore given by the difference between the two rates

$$r_{if} = r_{i \to f} - r_{f \to i} = \frac{2\pi}{\hbar} |\mathbf{H}_{if}|^2 \delta(E_f - E_i - E)(f_i - f_f). \tag{4.9}$$

For all such transitions involving a quantum of energy E we should sum over all pairs of initial and final states which differ by E. For a crystal, where states are distinguished by wavevector, we can convert the sum over discrete states $r(E) = \sum_{i,f} r_{if}$, into an integral over all initial states $\mathbf{k}_i$ and final states $\mathbf{k}_f$ which differ in energy by E, by introducing the density of states functions $g_f(\mathbf{k}_f)$ and $g_i(\mathbf{k}_i)$ for the relevant crystal bands (Eq. 3.11). Then we have for the transition rate per unit crystal volume,

$$r(E) = \frac{2\pi}{\hbar} \int\int |\mathbf{H}_{if}|^2 \delta(E_f - E_i - E)(f(E_i(\mathbf{k}_i)) - f(E_f(\mathbf{k}_f)))$$
$$\times g_i(\mathbf{k}_i) g_f(\mathbf{k}_f) d^3\mathbf{k}_i d^3\mathbf{k}_f \tag{4.10}$$

The form of $\mathbf{H}_{if}$ depends upon the nature of the interaction. For instance, for optical transitions the matrix element delivers the condition that $\mathbf{k} = \mathbf{k}'$, and the matrix element then depends only on E.

If the $\mathbf{k}$ dependence is not needed explicitly (for instance, if the band structure is isotropic) then the integral can be expressed in terms of energies,

$$r(E) = \frac{2\pi}{\hbar} \int |\mathbf{H}_{if}|^2 (f(E_i) - f(E_i + E)) g_i(E_i) g_f(E_i + E) dE_i \tag{4.11}$$

To find the *total* rate of band to band transitions, $r(E)$ should be summed over the transition energy E, weighted by the density of available photon states $g_{ph}(E)$ of the perturbing field

$$G = \int r(E) g_{ph}(E) dE \tag{4.12}$$

Optical transitions may involve additional photons, or phonons. In such cases the first order approximation given by Fermi's Golden Rule fails, and it is necessary to expand the quantum mechanical transition rate to higher order. An important example is the case of optical transitions in an indirect band gap semiconductor, which involve a photon and a phonon. The higher order terms result in a different energy dependence of the absorption coefficient for indirect and direct gap materials, which is discussed below. The higher order approximations are beyond the scope of this book.

4.3.2. *Optical processes in a two level system*

Here we consider the simple case of a steady state light field interacting with a two level system. The two level model allows us to derive the distribution function for photons, which will be used later. The photon energy is equal to the difference in energies E_v and E_c. Since the light biases the system the two levels will not be in thermal equilibrium, but we will assume that they are each in quasi thermal equilibrium with quasi Fermi levels of E_{F_p} and E_{F_n} respectively [de Vos, 1992].

The following events can occur:

- a photon can be absorbed to promote an electron from E_v to E_c;
- an electron can relax to E_v from E_c while emitting a photon (spontaneous emission);
- a photon can *stimulate* the relaxation of an electron from to E_v from E_c together with the emission of a second photon (stimulated emission) (Fig. 4.2).

The first and third of these processes are closely related. They can be distinguished *only* by the relative occupation probabilities of the two levels, and not by the strength of the photon field: any photon can cause either type of event. Therefore we will combine them into the *net* absorption with a rate given by

$$r_{abs} = \frac{2\pi}{\hbar}|\mathbf{H}_{cv}|^2 f \cdot (f_v - f_c) \tag{4.13}$$

where f_v is the (Fermi Dirac) probability that there is an electron is the state E_v,

$$f_v = \frac{1}{e^{(E_v - E_{F_p})/k_B T} + 1}, \tag{4.14}$$

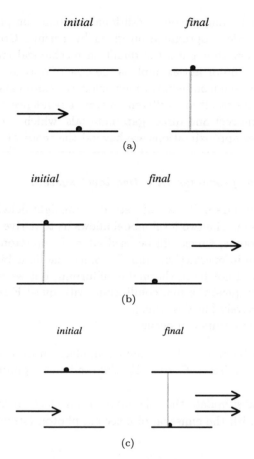

Fig. 4.2. (a) Absorption of a photon to promote an electron. (b) Relaxation of an electron accompanied by spontaneous photon emission. (c) Absorption of a photon to result in relaxation and stimulated photon emission.

f_c the probability that there is an electron in state E_c,

$$f_c = \frac{1}{e^{(E_c - E_{F_n})/k_B T} + 1},\tag{4.15}$$

f the probability that there is a photon, and $\mathbf{H}_{cv}$ is the optical matrix element.

Spontaneous emission proceeds at a rate

$$r_{sp} = \frac{2\pi}{\hbar}|\mathbf{H}_{cv}|^2 f_c \cdot (1 - f_v).\tag{4.16}$$

The same matrix element controls absorption and spontaneous emission because of microscopic reversibility: although the initial and final states in Fermi's Golden Rule have been exchanged, the value of H_{cv}^2 remains the same.

In the steady state, the rates of all three processes must balance. Setting

$$r_{abs} - r_{sp} = 0, \tag{4.17}$$

we find that

$$f = \frac{1}{e^{(E-\Delta\mu)/k_B T} - 1} \tag{4.18}$$

where we have used the definition of $\Delta\mu$

$$\Delta\mu = E_{F_n} - E_{F_p}. \tag{4.19}$$

It is clear from Eq. 4.19 that a more intense light field will cause a greater separation in quasi Fermi levels. For this reason, $\Delta\mu$ is sometimes referred to as the 'chemical potential of radiation' [Wuerfel, 1982].

f is the quasi-equilibrium occupation function for photons. Thus, the number density of photons of energy E in quasi thermal equilibrium is given by

$$n_{ph}(E) = f(E)g_{ph}(E) \tag{4.20}$$

where $g_{ph}(E)$ is the density of photon states. Away from equilibrium $n_{ph}(E)$ will be determined by the strength of the electromagnetic field.

Equation 4.17 also gives us the useful result that:

$$f_c(1 - f_v) = f \cdot (f_v - f_c). \tag{4.21}$$

4.4. Photogeneration

Photogeneration is by far the most important generation process in photovoltaic devices. By *photogeneration* we mean the generation of mobile electrons and holes through the absorption of light in the semiconductor. Other generation processes of relevance to photovoltaics are trap assisted and Auger generation, which will be mentioned in Chapters 8 and 10, respectively.

Though most relevant for photovoltaics, this is not the only optical process which occurs in semiconductors. Photons may alternatively be absorbed to increase the kinetic energy of mobile carriers (free carrier

absorption) or to generate phonons, or to promote electrons between localised states, or they may be scattered. The first two of these are usually important only at photon energies much smaller than the band gap
(< 100 meV) and the first only at high carrier densities. Near to the band
gap, band to band and localised state to band transitions are dominant.

Scattering, particularly by interfaces and by inhomogeneities in non
uniform media, removes light from the incident beam without generating
carriers, and so is undesirable for photovoltaics. However, it may be exploited in solar cell structures which are designed to trap the light and so
amplify the photon field. This will be discussed in Chapter 9.

4.4.1. *Photogeneration rate*

The macroscopic absorption coefficient α describes how the light intensity
is attenuated on passing through the material. $\alpha(E)$ may be considered as
the sum of the absorption cross sections per unit volume of material for
the various optical processes. Suppose a beam of photons of energy E and
intensity I_0 is normally incident on a slab of absorbing material. A fraction
$\alpha(E) \cdot dx$ of the photons of energy E entering a slab of thickness dx will
be absorbed and the light intensity $I(x)$ will be attenuated by a factor
$e^{-\alpha(E)dx}$. Hence

$$\frac{dI}{dx} = -\alpha I . \tag{4.22}$$

Integrating Eq. 4.22 for a material of non-uniform α the intensity at a depth
x, $I(x)$, is given by

$$I(x) = I(0)e^{-\int_0^x \alpha(E,x')dx'} \tag{4.23}$$

where I(0) is the intensity just inside the surface (*i.e.*, after accounting for
reflection). For uniform α, this reduces to the simple Beer–Lambert law,
$I(x) = I(0)e^{-\alpha x}$ (Fig. 4.3). It can be shown that α is related to the imagi-
nary part of the refractive index $\mathrm{Im}(n_s)$ through $\alpha = \frac{4\pi \mathrm{Im}(n_s)}{\lambda}$ where λ is the
wavelength of light. α defined this way may also contain contributions from
scattering; however, these are not considered in the microscopic derivation
below.

If we can assume that all photons are absorbed to generate free carriers
then the rate of carrier generation, per unit volume, at a depth x below the
surface is given by

$$g(E, x) = b(E, x)\alpha(E, x) \tag{4.24}$$

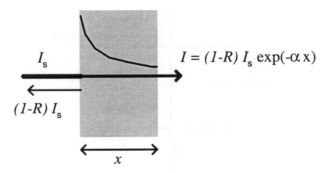

Fig. 4.3. Attenuation of light intensity in a slab of absorption α and thickness x. A fraction (1-R) of the light incident from the sun, I_s, is reflected at the front surface. The remaining intensity attenuates exponentially with distance in a uniform material, as shown by the thin black line. Photon flux density attenuates in the same way.

where b is the photon *flux* at x. Notice that it is the number and not the energy of the photons which determines the photogeneration rate. To relate g to the *incident* flux we need to allow for reflection of photons at the surface and attenuation within the material. Thus

$$g(E,x) = (1 - R(E))\alpha(E)b_s(E)e^{-\int_0^x \alpha(E,x')dx'} \tag{4.25}$$

where $b_s(E)$ is the incident flux and $R(E)$ is the reflectivity of the surface to normally incident light of energy E. This is the spectral photogeneration rate. To find the total generation rate at x we sum over photon energies

$$G(x) = \int g(E,x)dE. \tag{4.26}$$

The integral should be extended only over energies where photon absorption primarily results in free carrier generation.

4.4.2. *Thermalisation*

Photogeneration does not depend upon the energy of the absorbed photon, except in that the energy exceed the band gap. When higher energy photons are absorbed, they generate carriers with higher kinetic energy, but that energy is quickly lost and only E_g of potential energy remains to be collected, as shown in Fig. 4.4. The important quantity is the *number* of excitation events and not the amount of *energy* absorbed. This is one of the most important concepts in understanding photovoltaic devices.

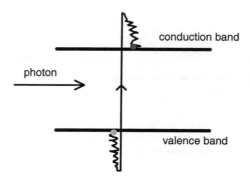

Fig. 4.4.   Photogeneration of an electron-hole pair and loss of excess electron and hole kinetic energy by thermalisation to the respective band edges.

It is also the essential difference between photovoltaic and solar thermal action.

The photogenerated carriers lose any extra kinetic energy by *thermalisation*, or *cooling*. Microscopically, this means that they undergo repeated collisions with the lattice, giving up some of their kinetic energy to produce a phonon while they decay into a lower energy state, until they are in thermal equilibrium with the ambient. Phonons are the means by which energy is carried away to the outside world. This cooling happens very fast, on the order of picoseconds, partly because of the high density of final states which are available at lower kinetic energy. Cooling is faster than all but the most intense generation processes. In exceptional circumstances, for example for very high optical intensities or for electronic structures with reduced density of available states, then the carriers may not be able to lose energy fast enough and consequently are 'hot'. This will be discussed in detail in Chapter 10.

4.4.3.   *Microscopic description of absorption*

Since photogeneration is the promotion of an electron from a valence to a conduction band energy level, the energy dependence of α must be strongly related to the density of valence and conduction band states and therefore to the band structure of the material. Here we will see how using Fermi's Golden Rule. Further details of the microscopic optical properties of semiconductors are given elsewhere [Bassani, 1975; Stern, 1963; Wuerfel, 1982; Bastard, 1986].

From Eq. 4.10, the net rate of transitions from an initial state in the valence band $|i\rangle = |v, \mathbf{k}_v\rangle$ to a final state in the conduction band $|f\rangle = |c, \mathbf{k}_c\rangle$ through an interval $E = E_c(\mathbf{k}_c) - E_v(\mathbf{k}_v)$ is given by

$$r(E) = \frac{2\pi}{\hbar} \iint |\mathbf{H}_{cv}|^2 \delta(E_c - E_v - E)(f_v(E_v(\mathbf{k}_v)) - f_c(E_c(\mathbf{k}_c)))$$

$$\times g_v(\mathbf{k}_v) g_c(\mathbf{k}_c) d^3\mathbf{k}_v d^3\mathbf{k}_c . \tag{4.27}$$

Now, for an electromagnetic field of strength E_0, polarisation vector ε and angular frequency ω, where the wavelength of light is long compared to interatomic distances, the perturbation $\mathbf{H}'$ is given in the dipole approximation by

$$\mathbf{H}' = \frac{iqE_0}{2m_0\omega} \varepsilon \cdot \mathbf{p} \tag{4.28}$$

where $\mathbf{p}$ is the quantum mechanical momentum operator. Then

$$r(E) = \frac{2\pi}{\hbar} \frac{q^2 E_0^2 \hbar^2}{4m_0^2 E^2} \iint M_{cv}^2 \delta(E_c - E_v - E)(f_v(E_v(\mathbf{k}_v)) - f_c(E_c(\mathbf{k}_c)))$$

$$\times g_c(\mathbf{k}') g_v(\mathbf{k}) d^3\mathbf{k} d^3\mathbf{k}' \tag{4.29}$$

where

$$M_{cv} = |\langle v, \mathbf{k}| \varepsilon \cdot \mathbf{p} | c, \mathbf{k}' \rangle| \tag{4.30}$$

is called the dipole matrix element.

Since $r(E)$ represents the net rate of photon absorption, the electromagnetic field is giving up *energy* to the semiconductor at a net rate

$$\frac{\partial U_E}{\partial t} = -E \times r(E) \tag{4.31}$$

per unit time, where U_E is the energy density of the radiation. Now, we use the fact that, for a plane wave, the cycle-averaged rate at which the beam loses energy with time is equal to the rate at which the beam loses intensity with distance,

$$\frac{\partial I}{\partial x} = \frac{\partial U_E}{\partial t} \tag{4.32}$$

if x is the direction of propagation. From Eq. 4.22 the absorption coefficient is given by

$$\alpha = -\frac{1}{I} \frac{dI}{dx} .$$

Now, since the intensity of radiation in a medium of refractive index n_s is related to U_E through

$$I = \frac{U_E c}{n_s} \tag{4.33}$$

and, from basic electromagnetic theory,

$$U_E = \frac{n_s^2 E_0^2}{8\pi} \tag{4.34}$$

we finally obtain for α:

$$\alpha = \frac{n_s}{c} \frac{E}{U_E} r_{abs}(E). \tag{4.35}$$

Substituting for $r_{abs}(E)$, from Eq. 4.29, we obtain

$$\alpha(E) = \frac{A_\alpha}{m_0} \iint M_{cv}^2 \delta(E_c - E_v - E)(f_v(E_v(\mathbf{k}_v)) - f_c(E_c(\mathbf{k}_c)))$$
$$\times g_c(\mathbf{k}')g_v(\mathbf{k})d^3\mathbf{k}d^3\mathbf{k}' \tag{4.36}$$

where

$$A_\alpha = \frac{4\pi^2 q^2 \hbar}{n_s c m_0 E}. \tag{4.37}$$

For low enough excitation levels, the valence band is effectively full ($f_v \approx 1$) and the conduction band empty ($f_c \approx 0$), so, to a good approximation

$$\alpha(E) = \frac{A_\alpha}{m_0} \iint M_{cv}^2 \delta(E_c - E_v - E)g_c(\mathbf{k}')g_v(\mathbf{k})d^3\mathbf{k}d^3\mathbf{k}'. \tag{4.38}$$

The macroscopic absorption coefficient may alternatively be considered as a sum of contributions from all valence (i) to conduction (f) band transitions which differ in energy by E

$$\alpha(E) = \sum_{i,f} \alpha_{if}(E) \tag{4.39}$$

where

$$\alpha_{if}(E) = \frac{A_\alpha}{m_0} M_{if}^2 \delta(E_f - E_i - E) \tag{4.40}$$

In this representation, the density of $\mathbf{k}$ states is included through the summation over different transitions.

Our expression for α will simplify when we evaluate the matrix element. At this stage it is worth distinguishing direct and indirect gap materials.

4.4.4. *Direct gap semiconductors*

In a crystal, electron and hole states can be written, using the Bloch theorem, as the product of a plane wave and a rapidly varying atomic part. Thus

$$|c, \mathbf{k}'\rangle = u_c(\mathbf{r})e^{i\mathbf{k}'\cdot\mathbf{r}} \tag{4.41}$$

and

$$|v, \mathbf{k}\rangle = u_v(\mathbf{r})e^{i\mathbf{k}\cdot\mathbf{r}}.$$

Now, acting on $|c, \mathbf{k}'\rangle$ with the differential operator $-i\hbar\varepsilon \cdot \nabla$, where the momentum operator has been written as $-i\hbar\nabla$, and writing the matrix element in integral notation we obtain for the matrix element

$$M_{cv} = -i\hbar\varepsilon \cdot \int [e^{i(\mathbf{k}'-\mathbf{k})\cdot\mathbf{r}}u_v^*(\mathbf{r})\nabla u_c(\mathbf{r}) + i\mathbf{k}'e^{i(\mathbf{k}'-\mathbf{k})\cdot\mathbf{r}}u_v^*(\mathbf{r})u_c(\mathbf{r})]d^3\mathbf{r}. \tag{4.42}$$

First, the integral is non zero only when $\mathbf{k}' - \mathbf{k} = 0$. In other words, momentum must be conserved. This is the usual *selection rule* for optical transitions. It is the reason for the distinction between direct and indirect gap materials. Second, the contribution from the second term in the intergrand must vanish since the different bands c, and v, must be orthoghonal. This is a consequence of quantum mechanics. Then the dipole matrix element is simply given by $\langle c|\varepsilon \cdot \mathbf{p}|v\rangle$. M_{cv} is a property of the material and, in general, of the electric field direction. For isotropic materials $M_{cv}^2 = \frac{1}{3}|\langle c|\mathbf{p}|v\rangle|^2$.

Now α can be evaluated simply from the densities of states. When these depend only on energy we have

$$\alpha(E) = \frac{A_\alpha}{m_0}M_{cv}^2 \int g_c(\mathbf{k}(E_i + E))g_v(\mathbf{k}(E_i))dE_i. \tag{4.43}$$

The quantity given by the integral is known as the joint density of states (JDOS). In the case of the parabolic band approximation (introduced in Chapter 3), where $g_v(E)$ varies like $(E_{v0} - E)^{1/2}$ and $g_c(E)$ varies like $(E - E_{c0})^{1/2}$, it is straightforward to show that

$$\alpha(E) = \alpha_0(E - E_g)^{1/2} \tag{4.44}$$

where α_0 is a material dependent constant, and $E_g = E_{c0} - E_{v0}$. Thus, the absorption coefficient reflects the shape of the individual densities of states. In general the JDOS of a direct gap material will look like the product of the conduction band and valence band density of states (DOS) functions.

4.4.5. *Indirect gap semiconductors*

In indirect gap materials, this is not the case. Equation 4.42 showed that optical transitions cannot occur unless $\mathbf{k}' - \mathbf{k} = 0$. Therefore an electron cannot be excited from the valence band maximum to the conduction band minimum in an indirect gap material simply by the absorption of a photon. What can happen, however, is that the electron can be excited simultaneously with the absorption or emission of a *phonon*. In the first case the ground state is a composite state of electron in the valence band plus a phonon, and the final state is a composite of an electron in the conduction band and no phonon:

$$|i\rangle = |v, \mathbf{k}; \omega_{\mathrm{p}}, \mathbf{k}_{\mathrm{p}}\rangle$$

$$|f\rangle = |c, \mathbf{k}'; 0\rangle$$

where the phonon has energy $E_{\mathrm{p}} = \hbar\omega_{\mathrm{p}}$ and momentum $\hbar\mathbf{k}_{\mathrm{p}}$. Conservation of energy requires that

$$E_{\mathrm{p}} = E_{\mathrm{c}}(\mathbf{k}') - E_{\mathrm{v}}(\mathbf{k}) - \hbar\omega. \tag{4.45}$$

Evaluation of the matrix element delivers the condition due to momentum conservation that $\mathbf{k}' - \mathbf{k}_{\mathrm{p}} - \mathbf{k} = 0$. Transitions from the valence band maximum to the conduction band minimum are allowed if a phonon of suitable $\mathbf{k}$ is available (Fig. 4.5). The sum over all initial and final states must now involve the probability density of phonons of energy E_{p} and wavevector $\mathbf{k}_{\mathrm{p}}$, $n_{\mathrm{p}}(E_{\mathrm{p}})$, as well as the JDOS of conduction and valence bands.

We will estimate the absorption coefficient for an indirect transition with phonon absorption. See also ([Pankove, 1971]). The transition rate

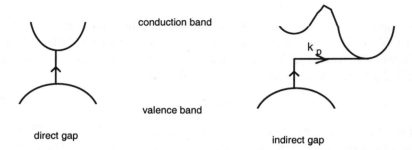

Fig. 4.5. Absorption in a direct and indirect gap band structure. In the indirect gap material, a phonon of momentum $\mathbf{k}_{\mathrm{p}}$ is needed to conserve momentum.

depends upon the availability of a phonon of suitable energy, the density of occupied valence band states and of unoccupied conduction band states. According to Bose–Einstein statistics the probability of finding a phonon of energy E_p is given by $\frac{1}{e^{E_p/k_BT}-1}$. If the various matrix elements are energy independent, the conduction band is initially empty and the valence band full, then the transition rate varies like

$$r \propto \frac{1}{e^{E_p/k_BT}-1} \int g_c(E_i + E + E_p)g_v(E_i)dE_i \qquad (4.46)$$

where the integral can be taken over energies, rather than **k**, since the phonon distribution is isotropic in **k**. Using the densities of states for parabolic bands such that $g_v(E) \propto (E_{v0} - E)^{1/2}$ and $g_c(E) \propto (E - E_{c0})^{1/2}$, the integral can be evaluated to find

$$\alpha_{p-} \propto \frac{(E_g - E - E_p)^2}{e^{E_p/k_BT}-1}.$$

A second type of indirect transition, where the photon absorption causes the *generation* of a phonon, should be included. In this case the transition rate is proportional to the probability of phonon emission, $\frac{1}{1-e^{-E_p/k_BT}}$ and the contribution to α varies like

$$\alpha_{p+} \propto \frac{(E_g - E - E_p)^2}{1 - e^{-E_p/k_BT}}.$$

The net absorption coefficient is the sum of contributions from each type of event. At photon energies which are such that $E - E_g \gg k_BT$, E_p can be neglected in the numerator and the absorption coefficient has the quadratic form

$$\alpha \propto (E - E_g)^2. \qquad (4.47)$$

The absorption coefficient of an indirect material has very different behaviour to a direct gap material and does not reflect the JDOS at photon energies close to the band edge. $\alpha(E)$ is generally smaller and rises more smoothly from the band edge than in direct gap materials. Similar behaviour is observed in highly doped semiconductors where electron scattering, rather than phonon emission and absorption, assists the optical transitions.

At higher photon energies, direct optical transitions are allowed in the indirect gap material. The contribution of these direct transitions should reflect the JDOS without any phonon contributions and so rise more rapidly with energy than the indirect contributions. The direct contributions usually appear as a change in curvature in $\alpha(E)$.

4.4.6.   *Other types of behaviour*

In practice, the $(E - E_g)^{1/2}$ behaviour predicted by Eq. 4.44 is seldom seen. At energies above E_g the DOS functions diverge from the parabolic approximation, while at energies just below E_g *excitons* influence the absorption.

The parabolic band approximation is only good close to the band extrema, typically within 100 meV. At values of $\mathbf{k}$ far from the band extrema the $E(\mathbf{k})$ curve begins to flatten out, and becomes stationary at other high symmetry values of $\mathbf{k}$. This leads to local maxima and other features in in the DOS, which can often be identified from peaks in the absorption spectrum.

As a better approximation to $\alpha(E)$ than $(E - E_g)^{1/2}$, parametric forms may be derived for groups of semiconductors sharing a similar crystal structure. These will depend typically upon the energies of the important symmetry points, band gaps and dipole matrix elements. Various models have been developed for the III–V and II–VI families of semiconductors [Adachi, 1992].

Multiple step photogeneration

As well as direct band to band photogeneration, free charge carriers may be generated indirectly, where the photon is absorbed to create an excited state which subsequently dissociates to release one or more free carriers. Examples are absorption by excitons and by sensitisers. Excitons are Coulombic bound states of electron-hole pairs, and were introduced in Chapter 3. An exciton can be created by a photon of energy smaller than E_g. The difference between the exciton energy and the band gap, called the exciton binding energy, is usually less than $k_B T$ and so most excitons dissociate at room temperature. However, the binding energy will be strong in cases where the photogenerated electron and hole wavefunctions are localised, as in the case of low dimensional semiconductor structures and molecular semiconductors. In those cases the excitons are important features of the room temperature absorption. The magnitude of the excitonic absorption, its oscillator strength, is generally larger for more strongly bound excitons. Note that only stationary ($\mathbf{k} = 0$) excitons can be generated optically.

The important point for application to photovoltaics is that excitonic absorption does not automatically generate mobile carriers. The exciton needs to dissociate first, and it may recombine before that happens. Therefore in modelling device behaviour we should use for G the *net* generation rate of free carriers from exciton generation and dissociation, and not simply the optical excitation rate.

Sensitisers are analogous to excitons and are usually deliberately introduced as optical absorbers. The sensitiser may be a molecular species such as a dye molecule, or a small solid state particle in contact with a semiconductor surface. Such systems are widely used in photography. The photon creates an excited state, which may then dissociate into a charged pair, following the sequence

$$S + h\nu \to S^* \Leftrightarrow S^+ + e^-$$

where S, S^* and S^+ represent the ground state, excited state, and ionised state of the sensitiser. The final stage involves injection of the free carrier into the semiconductor. Sensitisers at heterojunctions may inject charges of opposite sign into the two different media forming the junction [Hagfeldt, 2000].

In the case of absorption by excitons or sensitisers the optical generation rate should be replaced by

$$g(E, x) = (1 - R(E))\eta_{\text{diss}}(E)\alpha(E)b_{\text{s}}(E)e^{-\int_0^x \alpha(E, x')dx'}$$

where $\eta_{\text{diss}}(E)$ is the quantum efficiency for dissociation, *i.e.*, the probability that one absorbed photon of energy E will generate a free charge.

Amorphous materials

In amorphous (non-crystalline) material, the lack of long-range order means that crystal momentum need not be conserved in an optical transition. The band gap is always 'direct'. Absorption events are therefore more likely than in the equivalent crystalline material; for example, the absorption coefficient is larger in amorphous than in crystalline silicon. The form of $\alpha(E)$ is dominated by the JDOS without **k** conservation restrictions. This will be discussed for the case of amorphous silicon in Chapter 8.

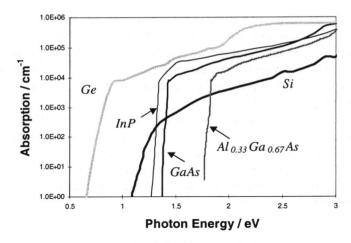

Fig. 4.6. Absorption spectra of some common photovoltaic semiconductors. Notice how the band edge of the direct semiconductors, GaAs, InP and $Al_xGa_{1-x}As$ is sharper than for the indirect semiconductors silicon and germanium. The sharp edges of the GaAs and InP absorption are influenced by excitonic effects.

4.4.7. *Examples and data*

Figure 4.6 shows the absorption spectra of a number of semiconductors which are important for photovoltaics. Notice how the absorption edge for the direct band gap semiconductors GaAs and InP, is sharper than for the indirect band gap materials, silicon and germanium. Notice how the shape of the curves for GaAs, InP and $Al_xGa_{1-x}As$ are similar; this is due to their similar crystal structure.

The absorption length of a photovoltaic material is a useful quantity. This is defined as the distance light of a particular wavelength must travel before the intensity is attenuated by a factor e, and is given by $\frac{1}{\alpha}$. At visible wavelengths the direct band gap materials GaAs and InP have absorption lengths of less than one micron. This means that only a few microns of material are needed to absorb virtually all of the light. In contrast, the indirect gap material Si has an absorption length of tens of microns, so that wafers tens or hundreds of μm thick are needed for good absorption.

Reflectivity has not been treated explicitly in this section. For typical semiconductors R(E) is $\approx$ 30–40% at visible wavelengths, and so it is an important factor. The net reflectivity of a semiconductor surface can be reduced using anti-reflection coatings or surface texturing; these are treated in Chapter 9.

4.5. Recombination

4.5.1. *Types of recombination*

By *recombination* we refer to the loss of mobile electrons or holes by any of a number of removal mechanisms. Unlike generation, where one mechanism is dominant, several different recombination mechanisms are important for the photovoltaic device.

We should distinguish two categories: *unavoidable* recombination processes which are due to the essential physical processes in the intrinsic material, and *avoidable* processes which are largely due to imperfect material.

Amongst the unavoidable recombination processes are the processes which result from optical generation, spontaneous and stimulated emission (see Sec. 4.3). For photovoltaics, the most important of these is spontaneous emission, which is also known as radiative recombination.

The other important unavoidable process is the interaction of an electron or hole with a second similar carrier, resulting in the decay of one carrier across the band gap and the increase in the kinetic energy of the other carrier by an amount equal to the band gap. This is called *Auger* recombination. It is the reverse of a rare generation process where a carrier with kinetic energy greater than the band gap is able to give up some of its kinetic energy to excite an electron across the gap (discussed in Chapter 10). Auger recombination is important in low band gap materials with high carrier densities, where carrier–carrier interactions are stronger.

Avoidable recombination processes usually involve relaxation by way of a localised trap state. These trap states are due to impurities in the crystal or defects in the crystal structure. These are often known as non-radiative recombination processes (although Auger is also non-radiative) and are usually the dominant mechanisms. Recombination in semiconductors is well covered by Landsberg [Landsberg, 1990] and many semiconductor textbooks such as Shur [Shur, 1990] or Tyagi [Tyagi, 1991].

4.5.2. *Radiative recombination*

Box 4.2. Derivation of the radiative recombination rate

Here we are going to derive an expression for the rate of radiative recombination using our results for absorption coefficient.

From Sec. 4.3 above we know that the rate of spontaneous relaxation events from an initial state $|f\rangle = |c, \mathbf{k}_c\rangle$ of energy E_c to a final state $|i\rangle =$

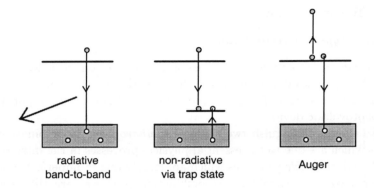

radiative
band-to-band

non-radiative
via trap state

Auger

Fig. 4.7.   Radiative band-to-band, non-radiative and Auger recombination.

$|v, \mathbf{k}_v\rangle$ of energy E_v is given by

$$r_{\mathrm{sp}} = \frac{2\pi}{\hbar}\mathbf{H}_{\mathrm{cv}}^2 f_c \cdot (1 - f_v)$$

(Eq. 4.13) while the net rate of absorption is

$$r_{\mathrm{abs}} = \frac{2\pi}{\hbar}\mathbf{H}_{\mathrm{cv}}^2 f \cdot (f_v - f_c)$$

(Eq. 4.16) where f is the probability that there is a photon of energy $E = (E_c - E_v)$. At quasi thermal equilibrium the two rates match and we obtain Eq. 4.18 for f. We will call this $f_{\mathrm{eq}}(E)$. Away from equilibrium

$$r_{\mathrm{sp}} = r_{\mathrm{abs}} \times \frac{f_{\mathrm{eq}}}{f} . \tag{4.48}$$

Now, under an incident field the probability that there is a photon of energy E is related to the (non-equilibrium) number density of photons n_{ph} by

$$f = \frac{n_{\mathrm{ph}}}{g_{\mathrm{ph}}} \tag{4.49}$$

where g_{ph} is the density of photon states in the energy range E to $E + dE$. In an optically isotropic medium of refractive index n_{s}

$$g_{\mathrm{ph}}(E) = \frac{8\pi n_{\mathrm{s}}^3 E^2}{h^3 c^3} \tag{4.50}$$

n_{ph} is related to the field energy density through

$$n_{\mathrm{ph}} = \frac{U_{\mathrm{E}}}{E} \tag{4.51}$$

Now we can relate the spontaneous emission rate to the absorption coefficient, using Eq. 4.35 to relate r_{abs} to α,

$$r_{sp} = \frac{r_{abs}E}{U_E} \times g_{ph}f_{eq} = \frac{c}{n_s}\alpha(E)g_{ph}(E)f_{eq}(E). \tag{4.52}$$

Substituting for g_{ph} and f_{eq} we have

$$r_{sp} = \frac{8\pi n_s^2}{h^3c^2}\frac{\alpha(E)E^2}{e^{(E-\Delta\mu)/k_BT}-1}. \tag{4.53}$$

This is the volume rate of radiative recombination, taken over all angles of emission. For applications in devices we are most often interested in the recombination resolved along a particular direction. In that case, the geometric factor is given by π rather than 4π and the recombination rate is reduced by a factor $1/4$:

$$r_{sp} = \frac{2\pi n_s^2}{h^3c^2}\frac{\alpha(E)E^2}{e^{(E-\Delta\mu)/k_BT}-1}. \tag{4.54}$$

This rate can be expressed in terms of the photon flux from a (biased) black body, $b(E,\Delta\mu)$

$$r_{sp} = b_e(E,\Delta\mu)\alpha(E) \tag{4.55}$$

where

$$b_e(E,\Delta\mu) = F_a\frac{2}{h^3c^2}\frac{E^2}{e^{(E-\Delta\mu)/k_BT}-1}$$

as shown in Eq. 2.14, where $F_a = \pi$ for emission normal to the cell surface and $n_s = 1$.

We obtain the total radiative recombination rate by integrating 4.55 over photon energies:

$$U_{rad}^{total} = \int_0^\infty \alpha(E)b_e(E,\Delta\mu)dE.$$

As explained in Sec. 4.1, to obtain the *net* recombination rate we must subtract the rate at thermal equilibrium, which is the rate when $\Delta\mu = 0$,

$$U_{rad} = \int_0^\infty \alpha(E)b_e(E,\Delta\mu)dE - \int_0^\infty \alpha(E)b_e(E,0)dE. \tag{4.56}$$

Spatial variations in the recombination rate enter through spatial variations in $\Delta\mu(x)$ (see also Wuerfel [Wuerfel, 1982] and Stern [Stern, 1963]).

4.5.3. *Simplified expressions for radiative recombination*

In order to use the recombination rate into the transport equations, it would be useful to be able to express U_{rad} in terms of the carrier densities. In a non-degenerate semiconductor, $E - \Delta\mu \gg k_B T$ for all energies E at which the rate, Eq. 4.55, is non-negligible. Then, to a good approximation,

$$\frac{E^2}{e^{(E-\Delta\mu)/k_B T} - 1} \approx e^{-(E-\Delta\mu)/k_B T} E^2 .$$

Using the relation for a semiconductor under bias, Eq. 3.53,

$$np = e^{\Delta\mu/k_B T} n_i^2 ,$$

we can write Eq. 4.56 as

$$U_{rad} = B_{rad}(np - n_i^2) \tag{4.57}$$

where

$$B_{rad} = \frac{1}{n_i^2} \frac{2\pi}{h^3 c^2} \int_0^\infty n_s^2 \alpha(E) e^{-E/k_B T} E^2 dE . \tag{4.58}$$

The radiative recombination coefficient B_{rad} is carrier density independent and is a property of the material.

The expression simplifies further for doped material. If the density of photogenerated electrons and holes is each equal to Δn, then in p type material with a doping density N_a,

$$(np - n_i^2) = (n_0 + \Delta n)(p_0 + \Delta n) - n_0 p_0 \approx \Delta n N_a$$

where n_0 and p_0 are the electron and hole densities in equilibrium, given by n_i^2/N_a and N_a, respectively. This means that U_{rad} is proportional to the excess minority carrier density,

$$U_{rad} = \frac{n - n_0}{\tau_{n,rad}} \tag{4.59}$$

where

$$\tau_{n,rad} = \frac{1}{B_{rad} N_a} \tag{4.60}$$

is the *minority carrier radiative lifetime*. A similar analysis shows that in n type material of doping density N_d,

$$U_{rad} = \frac{p - p_0}{\tau_{p,rad}} \tag{4.61}$$

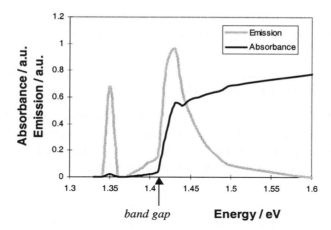

Fig. 4.8. Absorption and calculated emission spectrum for a slab of GaAs with a small density of defect levels below the band edge. The emission spectrum is strongly influenced by the shape of the absorption near the band edge. Notice how a small density of impurity states below the band edge, which is barely visible in the absorption spectrum, is strongly visible in the emission spectrum.

with

$$\tau_{\mathrm{p,rad}} = \frac{1}{B_{\mathrm{rad}}N_{\mathrm{d}}}. \tag{4.62}$$

The radiative lifetime can be measured from the time resolved spontaneous emission following instantaneous optical excitation of the semiconductor (called photoluminescence or fluorescence). B_{rad} can be determined experimentally from the variation of τ_{rad} with doping.

Equation 4.58 shows that B_{rad} is larger for materials with a high absorption coefficient, and therefore radiative recombination is more important in direct band gap materials. The exponential term means that, relative to absorption, contributions from energy levels closer to the band edges are much more important. It also means that radiative recombination from either band to impurity states inside the band gap can be very important, and can dominate over band-to-band events.

In cases where the chemical potential is uniform, the radiative recombination rate can be directly related to bias. If $\Delta\mu(x) \approx qV$ throughout the material then, using Eq. 4.57,

$$U_{\mathrm{rad}} = U_{\mathrm{rad,0}}(e^{qV/k_{\mathrm{B}}T} - 1) \tag{4.63}$$

where $U_{\mathrm{rad,0}}$ is a material dependent constant.

Radiative recombination is unimportant for practical cells at operating point but in the limit of perfect material it is the mechanism which limits efficiency.

Box 4.3. Bimolecular recombination

The detailed treatment of radiative recombination above shows how the rate of recombination is proportional to the densities of both types of carrier. In fact, it is usual for band-to-band recombination processes which do not depend on the presence of a third quantity to vary like

$$U = Bnp$$

where B is a material and process dependent constant. We show here that this is consistent with Fermi's Golden Rule.

For transitions from valence to conduction band, we may replace $(f_v - f_c)$ with $(1 - f_v) \cdot f_c$ in Eq. 4.27. Since under normal conditions the conduction band is mainly empty while the valence band is mainly full, the Fermi Dirac functions vary exponentially with energy over the relevant ranges of E_v and E_c. The factors f_c and f_v are therefore likely to be the most rapidly varying factors. If the matrix element is only weakly dependent on initial and final state energies, it can be taken outside the integral in the expression for the transition rate in and the integrations over the conduction and valence bands separate. In the case where the band structure is isotropic,

$$U = \frac{2\pi}{\hbar}|\mathbf{H}'_{cv}|^2 \int_{-\infty}^{E_{v,0}} (1 - f_v(E_v))g_v(E_v)dE_v \int_{E_{c0}}^{\infty} f_c(E_c)g_c(E_c)dE_c . \quad (4.64)$$

The two integrals in Eq. 4.64 are in fact the density of holes in the valence band and electrons in the conduction band. Thus, the transition rate is proportional to the product of the free carrier densities, $U \propto np$. (Remember that the thermal recombination rate must be subtracted to obtain the net recombination rate!)

For transitions from conduction band to a localised site, or from localised site to valence band, we replace the conduction (or valence) band DOS function with the DOS for the localised state, and n (or p) with the density of electrons (or holes) in the localised state. For conduction band to localised state transitions we have

$$U = BnN_t(1 - f_t) \quad (4.65)$$

where N_t is the density of localised states and f_t the probability that the localised state is occupied.

Relationships between rates and carrier densities are more complicated when the initial to final state transition is a multi step process and when the transition depends on the availability of a third entity. Shockley Read Hall recombination via a trap state in the band gap is an example of a multi-step process and Auger recombination, considered next, is a three carrier process.

4.5.4. *Auger recombination*

In *Auger* recombination, a collision between two similar carriers results in the excitation of one carrier to a higher kinetic energy, and the recombination of the other across the band gap with a carrier of opposite polarity. The energy which is released through recombination is given up as kinetic energy to the other carrier. Ultimately that extra energy will be lost as heat as the excited carrier relaxes to the band edge.

For band-to-band Auger recombination, an electron and two holes or a hole and two electrons are involved. By a similar argument to that used above, the rate is proportional to the densities of all three carriers, so that the net rate varies like

$$U_{\text{Aug}} = A_{\text{p}}(n^2 p - n_0^2 p_0)$$

for two-electron collisions and

$$U_{\text{Aug}} = A_{\text{n}}(np^2 - n_0 p_0^2) \tag{4.66}$$

for two-hole collisions. Auger processes are most important where carrier densities are high, for instance in low band gap and doped materials, or at high temperature. The dependence on doping density is strong. In p-type doped material the electron lifetime for band-to-band Auger recombination is given by

$$\tau_{\text{n,Aug}} = \frac{1}{A_{\text{n}} N_{\text{a}}^2}. \tag{4.67}$$

Similarly, the hole lifetime in n-type material is

$$\tau_{\text{p,Aug}} = \frac{1}{A N_{\text{d}}^2}.$$

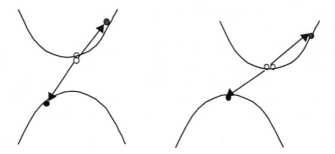

Fig. 4.9.   Energy-momentum representation of Auger recombination in direct (left) and indirect (right) band gap materials. A collision between two electrons near to the minimum of the conduction band results in the promotion of one electron to a higher energy state and recombination of the other with a hole. Both energy and momentum must be conserved. The right-hand figure shows that Auger events may occur between conduction band minimum and valence band maximum in indirect band gap materials, since the difference in momentum can be taken up by the promoted electron.

Auger recombination can also occur via a trap state. An electron colliding with an occupied trap state close to the conduction band can stimulate the recombination of the electron in the trap state with a valence band hole, whilst gaining kinetic energy. Similarly a hole colliding with an empty trap close to the valence band can stimulate Auger recombination. In these cases the minority carrier lifetimes vary like $\frac{1}{N_a N_t}$ and $\frac{1}{N_d N_t}$.

Auger events conserve momentum as well as energy. An electron with energy E and momentum $\mathbf{k}$ can recombine with a hole of energy $E - E'$ and momentum $\mathbf{k}'$, provided that there is an electron state available at $(E + E')$, $\mathbf{k} + \mathbf{k}'$ (see Fig. 4.9). This means that Auger recombination *can* occur in indirect band gap materials, unlike radiative recombination which is suppressed. Auger recombination is therefore much more important in indirect than direct band gap materials, and is the dominant loss mechanism in very pure silicon and germanium.

4.5.5.   *Shockley Read Hall recombination*

By far the most important recombination processes in real semiconductors are those which involve defect or *trap* states in the band gap. Since a trap state is spatially localised whilst the free electron or hole is delocalised, we can think of the free carrier as being captured by the trap. The carrier can subsequently be released by thermal activation. Alternatively, if the trap captures a carrier of the opposite polarity before the first carrier is released, then the trap has been emptied again and the two carriers have in effect recombined. This is illustrated in Fig. 4.11. Localised states which serve mainly to capture and release only one type of carrier are usually called traps. Those which capture both types of carrier are called recombination centres. Usually recombination centres lie deeper into the band gap than traps.

Box 4.4. Derivation of the Shockley Read Hall (SRH) recombination rate

Consider a semiconductor containing a density N_t trap states at an energy E_t in the band gap. Empty traps can capture electrons from the conduction band, and filled traps can capture holes from the valence band. The rate at which electrons are captured in this bimolecular process is given by Eq. 4.65

$$U_{nc} = B_n n N_t (1 - f_t)$$

where f_t is the probability that the trap is occupied. The coefficient can be expressed as

$$B_n = v_n \sigma_n \tag{4.68}$$

where v_n is the mean thermal velocity of the electron, and σ_n the capture cross section of the trap for electrons. It will also be useful to define a lifetime for electron capture by the trap

$$\tau_{n,\text{SRH}} = \frac{1}{B_n N_t}. \tag{4.69}$$

The rate at which electrons are released from the trap depends on the occupation of the traps and can be written

$$G_{nc} = \frac{N_t f_t}{\tau_{esc}} \tag{4.70}$$

where the release time τ_{esc} is determined by the condition that $U_{nc} = G_{nc}$ in equilibrium. Then it follows that

$$G_{nc} = B_n n_t N_t f_t \tag{4.71}$$

where n_t is the value of the electron density when the electron Fermi level is equal to the trap level,

$$n_t = n_i e^{(E_t - E_i)/k_B T}. \tag{4.72}$$

In a similar way, holes are captured at a rate

$$U_{pc} = B_p p N_t f_t \tag{4.73}$$

where

$$B_p = \nu_p \sigma_p \tag{4.74}$$

and

$$\tau_{p,SRH} = \frac{1}{B_p N_t}, \tag{4.75}$$

and released at a rate

$$G_{pc} = B_p p_t N_t (1 - f_t) \tag{4.76}$$

where p_t is the value of the hole density when $E_{F_p} = E_t$,

$$p_t = n_i e^{(E_i - E_t)/k_B T}. \tag{4.77}$$

In general, the rates of capture and release for electrons and holes will be different, depending on the affinity of the trap for electrons or holes and its position in the band gap. However, in the steady state, the net rate of electron capture by the traps $U_{nc} - G_{nc}$ must be equal to the net rate of hole capture $U_{pc} - G_{pc}$, since charges cannot be allowed to build up on the traps. This condition fixes the value of f_t,

$$f_f = \frac{B_n n - B_p p_t}{B_n(n + n_t) + B_p(p + p_t)}. \tag{4.78}$$

f_t can then be eliminated from the expressions for $U - G$. Finally we find for the net recombination rate

$$U_{SRH} = \frac{np - n_i^2}{\tau_{n,SRH}(p + p_t) + \tau_{p,SRH}(n + n_t)} \tag{4.79}$$

where we have used the definitions of $\tau_{n,SRH}$ and $\tau_{p,SRH}$. This is the Shockley Read Hall expression for recombination through a single trap state.

For doped semiconductors, U_{SRH} simplifies. In p type material, provided that $\tau_n N_a \gg \tau_p n_t$ and $N_a \gg p_t$, U_{SRH} becomes proportional to the excess carrier density,

$$U_{SRH} \approx \frac{(n - n_0)}{\tau_{n,SRH}}. \tag{4.80}$$

In n type material

$$U_{\mathrm{SRH}} \approx \frac{(p - p_0)}{\tau_{\mathrm{p,SRH}}}. \tag{4.81}$$

However, when n_{t} and p_{t}, or when $\tau_{\mathrm{n,SRH}}$ and $\tau_{\mathrm{p,SRH}}$ differ by orders of magnitude, these limiting forms may not apply. Then the full expression must be used.

SRH recombination is strongest when n and p are of similar magnitude. By examining Eq. 4.79 can see that for a mid gap trap with equal capture times, U has its maximum when $n = p$. (See Fig. 4.10.) This means that in undoped regions, where n and p may be similar, SRH recombination is more important relative to radiative recombination. (Equation 4.57 shows that radiative recombination depends only on the np product, which is constant for uniform $\Delta\mu$.) This dependence on the n/p ratio influences the bias dependence of the SRH recombination rate. We will see in Chapter 6 that SRH recombination varies like $e^{\mathrm{qV}/k_{\mathrm{B}}T}$ in doped material but like $e^{\mathrm{qV}/2k_{\mathrm{B}}T}$ in the 'depleted' layer between two differently doped materials.

In real semiconductors, there may be several trap levels in the band gap and carriers may recombine by more than one step. However, the largest contribution is from traps which are located close to the centre of the band gap, for $n \approx p$. So, for bulk material with a uniform density of traps, the approximation is usually made that recombination through one particular trap level is dominant. Multi-level recombination is discussed by Landsberg [Landsberg, 1990].

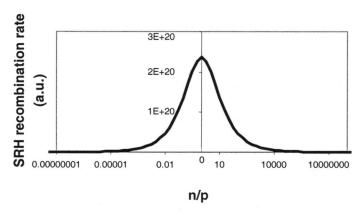

Fig. 4.10. SRH recombination rate as a function of the ratio n/p for a trap at mid-gap.

4.5.6.   *Surface and grain boundary recombination*

According to Eq. 4.79, spatial variations in U_{SRH} can be caused by variations in n and p, or by spatial variations in the nature or number density of the dominant trap. A higher density of trap states shortens the electron and hole capture times. In real materials, defects are much more likely to occur at surfaces and at the interfaces between different crystal regions in a multicrystalline or heterostructured material. Localised states at surfaces and interfaces include both crystal defects due to broken bonds, and extrinsic impurities which are deposited from the external environment, or which are concentrated at interfaces during growth. In such cases the trap states responsible for recombination are concentrated in a two-dimensional rather than three-dimensional space and it is much more meaningful to express the recombination in terms of the trap density per unit area of the surface or interface, than per unit volume. The relevant quantity will be a recombination flux — the number of carriers recombining at the interface per unit area per unit time, rather than a volume recombination rate.

If a surface contains a density N_s traps per unit area, then within a very thin layer δ_x around the surface, the surface recombination flux will be

$$U_s \delta x = \frac{n_s p_s - n_i^2}{\frac{1}{S_n}(p_s + p_t) + \frac{1}{S_p}(n_s + n_t)} \tag{4.82}$$

per unit area, where n_s, p_s are the electron and hole densities at the surface. S_n is the *surface recombination velocity* for electrons, defined by,

$$S_n = B_n N_s \tag{4.83}$$

and S_p the surface recombination velocity for holes,

$$S_p = B_p N_s . \tag{4.84}$$

In this definition, both S_n and S_p are directed towards the surface from the bulk.

In p type material, Eq. 4.82 reduces to

$$U_s \delta x \approx S_n (n_s - n_0) .$$

This leakage of minority carriers to the surface results in a *surface recombination current*. The magnitude of the current can be obtained from the electron continuity equation. In the dark, at steady state, Eq. 4.1 requires that $\nabla \cdot J_n = q U_n$. Integrating this across the interface layer, we find that the electron current density has changed by

$$\Delta J = J\left(x_s + \frac{1}{2}\delta x\right) - J\left(x_s - \frac{1}{2}\delta x\right) = q \int_{x_s-\delta x}^{x_s+\delta x} U\,dx = qS_n(n_s - n_0).$$

If the interface is a surface, then this condition determines the current density *at* the surface

$$J_n(x_s) = -qS_n(n_s - n_0).\tag{4.85}$$

Similarly, the change in hole current density at an interface in n type material is given by

$$\Delta J_p = -q \int_{x_s-\delta x}^{x_s+\delta x} U\,dx = -qS_p(p_s - p_0),$$

and the current density at the surface by

$$J_p(x_s) = qS_p(p_s - p_0).\tag{4.86}$$

4.5.7. *Traps versus recombination centres*

When a carrier is captured by a trap, it may then be released or it may be annihilated by the capture of the opposite type of carrier. When the time for electron release by thermal activation

$$\frac{1}{\tau_{esc}} = B_n n_i e^{(E_t - E_i)/k_B T}$$

is much shorter than the time for capture of a hole

$$\frac{1}{\tau_{cp}} = B_p p,$$

the state can be considered an *electron trap* rather than a recombination centre. This may happen if the state is close in energy to the conduction band edge, or if the cross section for electron capture is much larger than for hole capture, as, for instance, for a positively polarised defect. Similarly, localised states which are close to the valence band or which have a higher cross section for hole capture act as *hole traps*. Traps serve to slow down the transport of carriers but they do not remove them.

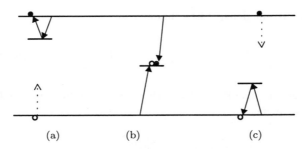

Fig. 4.11. (a) Electron trapping and detrapping; (b) electron-hole recombination; (c) hole trapping and detrapping.

4.6.   Formulation of the Transport Problem

Finally we can proceed to solve the set of equations for n, p and ϕ set out in the introduction to this chapter. In one dimension,

$$\frac{\partial n}{\partial t} = \frac{1}{q}\frac{\partial J_n}{\partial x} + G_n - U_n \tag{4.87}$$

$$\frac{\partial p}{\partial t} = -\frac{1}{q}\frac{\partial J_p}{\partial x} + G_p - U_p \tag{4.88}$$

and

$$\frac{d^2\phi}{dx^2} = \frac{q}{\varepsilon_s}\left(-\rho_{\text{fixed}}(x) + n - p\right). \tag{4.89}$$

For G_n and G_p we provide the photogeneration rate at that point. For band-to-band generation, $G = G_n = G_p$ is given by Eqs. 4.25 and 4.26. Photogeneration is the only generation process normally considered explicitly for solar cells; *thermal* generation of carriers is taken into account in each of the expressions for recombination. To evaluate $G(x)$ we need to know the absorption coefficient $\alpha(E, x)$, the reflectivity $R(E)$, and the incident photon flux density $b_s(E)$.

For U_n or U_p we should, in general, provide the sum of all the recombination processes — radiative, Auger, and trap assisted — as a function of n, p, ϕ and x. For radiative recombination we also need to know the coefficient B_{rad} (Eqs. 4.57 and 4.58). For Auger recombination we need the Auger coefficients (Eq. 4.66). For SRH recombination in the bulk we need the density and position of the dominant trap state and the lifetimes for electron and hole capture (Eq. 4.79); at the surface we need the recombination velocities (Eqs. 4.83 and 4.84).

For the fixed charge density in Poisson's equation we need the doping profile

$$\rho_{\text{fixed}}(x) = (-N_a(x) + N_d(x)) \tag{4.90}$$

where N_a and N_d represent the densities of *ionised* acceptors and donors.

J_n and J_p are derived for a crystalline material in Chapter 3. According to Eqs. 3.65 and 3.66, the currents can be expressed in terms of the electron and hole quasi Fermi levels, in one dimension,

$$J_n = \mu_n n \frac{dE_{F_n}}{dx}$$

$$J_p = \mu_p p \frac{dE_{F_p}}{dx} .$$

For a crystalline material, these can be expressed in terms of n and p using the definitions of E_{F_n}, E_{F_p} (Eqs. 3.31 and 3.34). For non-crystalline materials we must substitute alternative appropriate expressions for J_n and J_p such as Eq. 3.81.

When J_n and J_p are substituted into Eqs. 4.87 and 4.88, we have a set of three equations which can be solved to deliver n, p and ϕ as a function of x and t assuming that all material parameters are known. Boundary conditions are provided by the external electrical conditions and by the surface recombination conditions, and by the time dependence of the electrical and optical conditions for the transient problem.

4.6.1. *Comments on the transport problem*

Steady state solutions

Since solar cells operate in the steady state we are usually interested in the case where

$$\frac{\partial n}{\partial t} = \frac{\partial p}{\partial t} = 0$$

and solve the transport equations for a steady state illumination and electrical conditions. In the steady state, the electron and hole densities in each band and in localised states must be constant. This means that the generation and capture terms for the exchange of carriers between band and *trap* states (as opposed to recombination centres) must cancel out, and so trapping can be left out of the continuity equations. The consequence is that the generation and recombination processes are all effectively band-to-band: $G_n = G_p$ and $U_n = U_p$. The trapped charge density will influence the solution through the fixed charge term in Poisson's equation.

In the transient case where traps are being filled or emptied, then the terms for capture and release from traps must be included and net generation rate for holes and electrons are nonzero.

Photon continuity

Until now, we have assumed that the density of photons at a point is determined solely by the absorption within the material. That is, we have assumed that the density of photons resulting from radiative recombination is negligible compared to the incident flux density. In materials with a high radiative efficiency, or under high illumination conditions, these 'recycled' photons may be significant. Then a further continuity equation is required, for photons, and a further unknown needs to be found, the photon flux density $b(x)$. The case of photon recycling is treated in Chapter 9.

4.6.2. *Transport equations in a crystal*

The most relevant context for conventional photovoltaics is a one-dimensionally varying crystalline material in the steady state. The quantity of interest is the net current at either of the terminals, $J_n + J_p$, and its dependence on applied bias and illumination. To find this we first need to set up the transport equations and solve for n, p and ϕ, as described above.

We proceed by combining the current equations for a compositionally invariant crystal (Eqs. 3.75 and 3.76) with the steady state continuity equations to obtain a pair of second order differential equations governing n and p. In one dimension

$$D_n \frac{d^2 n}{dx^2} + \mu_n F \frac{dn}{dx} + \mu_n n \frac{dF}{dx} - U + G = 0 \qquad (4.91)$$

and

$$D_p \frac{d^2 p}{dx^2} + \mu_p F \frac{dp}{dx} + \mu_p p \frac{dF}{dx} - U + G = 0. \qquad (4.92)$$

While G is usually a function of position only, U generally depends upon both carrier densities. This couples the transport equations for electrons and holes. In certain situations, however, U depends on n or p only and the equations can be solved independently. In particular, when

(i) one carrier type greatly exceeds the other and the recombination rate simplifies to the monomolecular form $U \approx (n - n_0)/\tau_n$ (Eq. 4.80), and

(ii) the electric field F is zero or constant, the transport equations for minority carriers simplify to the analytically soluble form:

$$\frac{d^2n}{dx^2} + \frac{qF}{k_BT}\frac{dn}{dx} - \frac{(n - n_0)}{L_n^2} + \frac{G(x)}{D_n} = 0 \qquad (4.93)$$

for electrons in the p region, and

$$\frac{d^2p}{dx^2} + \frac{qF}{k_BT}\frac{dp}{dx} - \frac{(p - p_0)}{L_p^2} + \frac{G(x)}{D_p} = 0 \qquad (4.94)$$

and for holes in the n region.

Here we have used the Einstein relation $\mu = \frac{qD}{k_BT}$ (Eq. 3.77) for μ and have introduced the *diffusion length* for electrons

$$L_n = \sqrt{\tau_n D_n} \qquad (4.95)$$

and holes

$$L_p = \sqrt{\tau_p D_p}. \qquad (4.96)$$

The diffusion length is a measure of the average distance a minority carrier will diffuse before recombining. In steady state, the one-dimensional diffusion equation becomes

$$\frac{(n - n_0)}{\tau_n} = D_n \frac{d^2n}{dx^2} \qquad (4.97)$$

which has solutions for the *excess* minority carrier concentrations $(n - n_0)$ of the form $e^{\pm x/\sqrt{D\tau}}$. This makes $L = \sqrt{D\tau}$ a natural unit of length to characterise diffusion. Minority carrier concentration and current profiles have different behaviour depending on whether L is large or small compared to the layer width, and to the absorption length.

Finally, we solve the transport equations for $(n - n_0)$, $(p - p_0)$ subject to appropriate boundary conditions. In the next chapters, we will meet several examples where the simplified approach of Eqs. 4.93 and 4.94 can be used.

4.7. Summary

Photocurrent generation by a solar cell is linked to charge carrier generation and recombination by conservation of the numbers of electrons and holes. Photogeneration is the primary carrier generation process in photovoltaic cells. For direct gap semiconductors, the absorption coefficient can be described mathematically by Fermi's Golden Rule, and near to the band edge it depends on photon energy approximately as $(E - E_g)^{1/2}$. For indirect gap materials, light absorption requires phonon absorption or emission, the absorption coefficient is generally smaller and rises more gradually, like $(E - E_g)^2$, near the band edge. Photogeneration formally requires the dissociation of an excited state and is not identical to light absorption in some compound or organic materials where this dissociation is incomplete.

The main recombination mechanisms are radiative, Auger, and trap assisted recombination in the bulk or at the surface. Radiative recombination is the relaxation of an electron across the band gap together with the emission of a photon. It is unavoidable in a light absorbing material and is most important when absorption is strong, in direct gap semiconductors. The radiative recombination rate is described by a generalised Planck formula and varies approximately as np. Auger recombination is the relaxation of a charge carrier to excite a second carrier to a higher energy state within the band. The rate depends on carrier densities to third order and is strongest when the charge carrier densities are high. Trap assisted recombination is a multiple step relaxation process, usually involving intermediate states in the band gap. The energy lost by relaxation is given up as heat. The rate depends upon the density and position of these intermediate states and on the relative densities of electrons and holes. It is strongest for deep traps when n and p are similar. In doped material it becomes linear and can be characterised by a single recombination time. Trap assisted recombination is particularly important at a surface on account of surface states. Surface recombination becomes linear in doped material and can be characterised by a surface recombination velocity which depends on the density of surface defects. In real materials trap assisted recombination is dominant. In the limit of perfect material, radiative recombination is dominant and Auger becomes important for low band gap and indirect gap materials.

Formulae for the generation and recombination rates can be used to set up transport equations for each of the charge carriers. Together with Poisson's equation, these form a set of coupled differential equations which can be solved for the charge carrier densities, currents and the electrostatic potential. In general, the problem is complex but can be greatly simplified through assuming a linear recombination rate and neglecting electric field, conditions which are valid in many device structures.

References

S. Adachi, *Physical Properties of III-V Semiconductor Compounds* (Chichester: Wiley, 1992).

F. Bassani and G. Pastori Parravicini, *Electronic States and Optical Transitions in Solids* (Oxford: Pergamon, 1975).

G. Bastard, *Wave Mechanics Applied to Semiconductor Heterostructures* (Editions de Physique, 1986)

A. de Vos, *Endoreversible Thermodynamics of Solar Energy Conversion* (Oxford University Press, 1992).

S.J. Fonash, *Solar Cell Device Physics* (New York, London: Academic, 1980).

M.A. Green, *Silicon Solar Cells: Advanced Principles and Practice* (Sydney: Centre for Photovoltaic Engineering, 1995).

A. Hagfeldt and M. Grätzel, "Molecular photovoltaics", *Acc. Chem. Res.* **33**, 269–277 (2000).

P.T. Landsberg, *Recombination in Semiconductors* (New York: Cambridge University Press, 1991).

J.I. Pankove, *Optical Processes in Semiconductors* (Englewood Cliffs: Prentice Hall, 1971).

M. Shur, *Physics of Semiconductor Devices* (Englewood Cliffs: Prentice Hall, 1990).

F. Stern, "Elementary theory of the optical properties of solids", *Solid State Physics* **15**, 300 (1963)

M.S. Tyagi, *Introduction to Semiconductor Materials and Devices* (Chichester: Wiley, 1991).

P. Wuerfel, "The chemical potential of radiation", *J. Phys. C — Solid State Physics* **15**, 3967–3985 (1982).

Chapter 5

Junctions

5.1. Introduction

Photovoltaic energy conversion results from charge generation, charge separation and charge transport. In Chapters 3 and 4, the processes of photogeneration and charge transport in a semiconductor were discussed. The remaining stage, charge separation, requires some kind of driving force. This driving force is absolutely key for photovoltaic energy conversion and must be built in to our device. In the language of Chapter 3, the driving force can be equated to a light-induced gradient in the quasi Fermi levels for electrons and holes. In an alternative, electrical picture, the light absorbing material is connected to the external circuit by paths of different resistance: one which has much lower resistance for negative than positive charge, and the other which has much lower resistance for positive than negative (Fig. 5.1).

In principle, there are many ways of providing a charge separating mechanism. It is normally provided by spatial variations in the electronic environment. In a crystalline semiconductor device, a junction between two

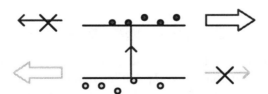

Fig. 5.1. Schematic of charge separation mechanism in a photovoltaic device. One contact provides a low resistance path for electrons but blocks hole flow, while the other provides an easy path for holes but a barrier to electrons. Such asymmetry can be achieved by using different electronic materials on either side.

electronically different materials provides an electrostatic force. In photosynthesis — where charge separation is also a requisite — excited electrons are driven across the photosynthetic membrane by differences in the free energy of molecular acceptors. In both cases the force arises from a compositional gradient.

In this chapter we will discuss, without mathematical detail, various types of junction used in photovoltaic devices. Chapter 6 completes the description of photovoltaic action with a detailed analysis of the most important example, the *p–n* junction.

5.2.   Origin of Photovoltaic Action

In a photovoltaic device, light produces a separation of charges. That charge separation then gives rise to a photocurrent (in short circuit) or a photovoltage (in open circuit). The photovoltaic *action* arises from the driving force separating charges. Here we will analyse this in terms of the contributions to the photocurrent, *J*. (It can equally well be analysed in terms of contributions to the photovoltage [Fonash, 1980].)

Now, from Chapter 3 (Eqs. 3.65–3.67) we have for the current at a point

$$J = J_n + J_p = \mu_n n \nabla E_{F_n} + \mu_p p \nabla E_{F_p} . \tag{5.1}$$

By definition, for a semiconductor in equilibrium, E_{F_n} and E_{F_p} are equal and constant and $J = 0$ everywhere. So, to achieve photovoltaic action, we must have a situation where light produces a gradient in at least one of the quasi Fermi levels.

How could that happen? Using the drift diffusion forms for J_n and J_p from Chapter 3 (Eqs. 3.75–3.76) we have that

$$J_n = qD_n \nabla(n - n_0) + \mu_n(n - n_0)(qF - \nabla\chi - kT\nabla \ln N_c) \tag{5.2}$$

and

$$J_p = -qD_p \nabla(p - p_0) + \mu_p(p - p_0)(qF - \nabla\chi - \nabla E_g + kT\nabla \ln N_v) \tag{5.3}$$

where n_0, p_0 represent the carrier densities in equilibrium and we have made use of the fact that $J_n = J_p = 0$ in equilibrium. In either equation, the first term represents *diffusion* and the second represents *drift* under the net electric field, which is due to compositional gradients, as well as any electrostatic field (as shown in Fig. 5.2). When the equations are written in this form, both terms are identically zero in equilibrium.

Under illumination, $n > n_0$ and $p > p_0$. Then, if the electric field is non zero, a net drift current will result. Alternatively, if there is no electric field, but there are gradients in the carrier densities, then a net diffusion current may result.

An *electric field* which exists in equilibrium is called a 'built-in' field and is due to compositional factors. We shall see below how a built-in electrostatic field is established at the interface between two materials of different work function. As shown in Chapter 3, an *effective* electric field may also result from gradients in the electron affinity, band gap, and effective band density of states. An electric field is effective for charge separation since it always drives p and n carriers in opposite directions.

Carrier density gradients result from gradients in the generation or removal rate. In an otherwise isotropic environment a gradient in the generation rate of electron-hole pairs can result in a *net* current only if the diffusion constants for electrons and holes are different. If they are the same, then the electron and hole diffusion currents cancel out exactly. The potential difference created by asymmetry in diffusion constants, the Dember potential, is not usually large enough for effective photovoltaic action in crystalline materials, although it may be large in molecular materials.

Large diffusive currents can be achieved, however, in an asymmetric environment where there is some additional mechanism which selectively remove electrons or holes. This could be a contact which has low resistance for electrons and high for holes. Such a contact or charged region preferentially removes electrons from the light absorbing region and creates a gradient in the electron density. It may be considered as an electron sink. That density gradient drives an electron diffusion current. In this situation there is no hole current to cancel the electron currents since holes were not collected. There may be a second contact which preferentially removes holes, and this creates a hole diffusion current which adds to the electron current.

In the following, we consider the contributions to the charge separating field.

Box 5.1. Contributions to the charge separating field

Figure 5.2 shows a band profile representation of the various possible contributions to the electric field experienced by electrons in the conduction band. (Note that each of these configurations results in an electron current

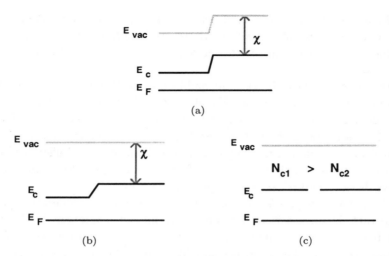

Fig. 5.2. Contributions to a built-in effective field for electrons. (See text for explanation.)

from right to left. The sign of the net current also depends upon the sense of the driving force for holes in the valence band.)

In (a) a difference in the work function has given rise to a gradient in the vacuum level and hence an electrostatic field, $\frac{1}{q}\nabla E_{\text{vac}}$.

In (b) a difference in the electron affinity due to a compositional gradient creates an effective field, $-\frac{1}{q}\nabla\chi$, seen as a gradient in the conduction band edge.

In (c) a field due to a gradient in the effective conduction band density of states, $-\frac{kT}{q}\ln\nabla N_{\text{c}}$, is driving electrons to the right. This term cannot be depicted on this diagram as it represents a gradient in the *free* energy rather than potential energy: carriers are driven thermodynamically in the direction of increasing availability of states.

Figure 5.3 shows the various factors which contribute to the net electric field within some general, compositionally varying semiconductor in equilibrium. Neglecting gradients in the band densities of states, an electron in the conduction band experiences an electric field of $\frac{1}{q}\nabla E_{\text{c}} = \frac{1}{q}(\nabla E_{\text{vac}} - \nabla\chi)$ and a hole in the valence band experiences the field $\frac{1}{q}\nabla E_{\text{v}} = \frac{1}{q}(\nabla E_{\text{vac}} - \nabla\chi - \nabla E_{\text{g}})$.

In Fig. 5.3(a) the carrier densities have their equilibrium profiles n_0, p_0, which are spatially varying, and the Fermi level is flat. In Fig. 5.3(b)

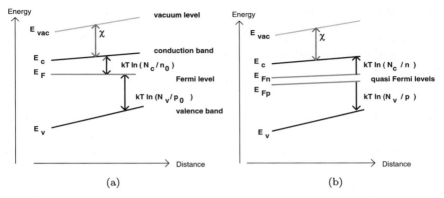

Fig. 5.3. Built in electric field and charge separation in a general material, (a) in equilibrium (b) under illumination.

a uniform excess carrier density of Δn is present so that $n_0 \rightarrow n_0 + \Delta n$ and $p_0 \rightarrow p_0 + \Delta n$. Now, the quasi Fermi levels are split and have both developed a positive gradient. Electrons will be driven to the left and holes to the right by the built-in field, thus effecting charge separation.

In summary, the various conditions which can give rise to charge separation in our semiconductor are:

(i) gradient in the vacuum level or work function $\Rightarrow$ electrostatic field
(ii) gradient in the electron affinity $\Rightarrow$ effective field
(iii) gradient in the band gap $\Rightarrow$ effective field
(iv) gradient in the band densities of states $\Rightarrow$ effective field

The first three of these are exploited in photovoltaic devices. Changes in (i), (ii) and (iii) can be achieved at the interface between two different materials — a 'heterojunction' — or through gradual changes in the composition of an alloy. However, the fields available through variations in the electron affinity or band gap can be fairly limited, at least in crystalline semiconductors. A further problem with heterojunctions is that defects at the interface can enhance recombination losses.

Changes in the vacuum level or work function (i), can also be achieved without using different materials, simply by varying the doping level in a single semiconductor. The fields established this way can be large and avoid the material problems due to heterojunctions. This is the most widely used means of establishing the charge separating field in solar cells.

The most common types of junction are described below.

5.3.   Work Function and Types of Junction

The work function of a material is the potential required to remove the least tightly bound electron. It is is defined by

$$\Phi_{\mathrm{w}} = (E_{\mathrm{vac}} - E_{\mathrm{F}}) \tag{5.4}$$

In metals Φ_{w} is always equal to the electron affinity. In semiconductors it can be controlled by doping. Since the position of the Fermi level within the band gap of a semiconductor depends upon doping, a particular semiconductor will have a smaller work function when it is doped n type than when it is doped p type.

Since E_{F} is constant in equilibrium, an electrostatic field must be established at the junction between any two regions of different Φ_{w}, due to the gradient in E_{vac}. This could be at a *heterojunction* between a semiconductor and a metal or two semiconductors of different Φ_{w}, or it could be at a *homojunction* between two layers of the *same* semiconductor which have been doped differently.

For any of these junctions, the electric field is evaluated in the following way:

The electrostatic potential energy difference across the junction is equal to the difference in the work functions. Therefore the field F established across a junction in the x direction must obey

$$q \int_{x_-}^{x_+} F \, dx = \Phi_{\mathrm{w}}(x_+) - \Phi_{\mathrm{w}}(x_-) = \Delta\Phi_{\mathrm{w}} \tag{5.5}$$

where x_- and x_+ represent positions far from the junction on either side where the electric field is zero. Poisson's equation requires that

$$\frac{d}{dx}(\varepsilon_{\mathrm{s}} F) = q\rho(x) \tag{5.6}$$

where ε_{s} is the local dielectric permittivity and the local charge density ρ includes contributions from all charges in the junctions region — trapped charges as well as fixed space charge and free electrons and holes. The free carrier distributions themselves depend on the local value of E_{F} relative to the band edges, so the problem has to be solved self consistently. Thus, accommodating a difference in work functions effectively means a redistribution of charge around the junction. This will be seen for the examples below.

5.4. Metal–Semiconductor Junction

Perhaps the simplest type of charge separating junction is the semiconductor–metal junction. We will use this as a model to explain in qualitative terms how the electrostatic field is established and how it causes photovoltaic action.

5.4.1. *Establishing a field*

Suppose we have a uniform n type semiconductor of work function Φ_n and a metal of work function Φ_m, such that $\Phi_m > \Phi_n$. When the two materials are isolated from each other, the Fermi levels will be independent, as in Fig. 5.4(a). When they are brought into electronic contact the Fermi levels must line up, as in Fig. 5.4(b), with the consequence that the vacuum level changes by $(\Phi_m - \Phi_n)$ between the semiconductor and the metal. Physically this is achieved by the exchange of charge carriers across the junction. Electrons flow from the semiconductor to the metal leaving a layer of positive fixed charge behind, and a negative image charge on the metal, until the charge gradient which builds up is sufficient to prevent further flow. At this point the two layers are in thermal equilibrium. The energy at the conduction band edge in the bulk of the semiconductor is lower than at the interface with the metal, and an electrostatic field exists close to the junction. The variation in electrostatic potential energy is represented by the change in E_{vac} in Fig. 5.4 and the field by the gradient of E_{vac}. The potential difference is shared between the two materials according to their

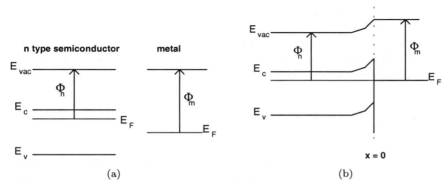

Fig. 5.4. (a) Band profiles of n-type semiconductor and metal in isolation. (b) Band profile of the semiconductor–metal junction in equilibrium.

dielectric permittivities. Because metals are much poorer at storing charge than semiconductors (much lower ε_s), virtually all of the potential difference is dropped in the semiconductor. At some distance from the junction on either side, the potential difference stops varying and the electric field falls to zero. This distance is vanishingly short in the metal but is significant — typically around a micron — in the semiconductor. Within these regions the materials carry a net charge. This region is called the *space charge region* or *space charge layer* of the junction. It corresponds to the region where E_{vac} is changing. Because the electron affinity and band gap are invariant in the semiconductor, the conduction and valence band energies must change in parallel with E_{vac}. This is usually referred to as *band bending*. The total amount by which the bands bend in the semiconductor is given by qV_{bi} where V_{bi} is called the built-in bias.

The charge distribution across the junction can be inferred from the band bending. Far from the junction, n and p will have their equilibrium values for that doping level, $n \approx N_d$, and $p \approx \frac{n_i^2}{N_d}$ (Eqs. 3.42 and 3.43) and the material is electrically neutral. Now since from Eq. 3.31 $n = N_c e^{-(E_c - E_F)/k_B T}$, we can see that as x approaches the junction and E_c increases, n will decrease below its equilibrium value so that $n < N_d$ and the material becomes positively charged. At the same time, the mobile hole concentration increases, since E_v is increasing and $p = N_v e^{-(E_F - E_v)/kT}$. However, p is still very small and the positive charge acquired is due mainly to the fixed ionised donor atoms which have lost their electrons. The space charge region is depleted of carriers and is often called a *depletion region* or *exhaustion layer*.

Thus, by joining an n type semiconductor to a metal of larger work function, we set up an electric field in a layer close to the interface. It is evident that this field will drive electrons to the left and holes to the right, so effecting charge separation. The contact presents a lower resistance path for holes than electrons, from semiconductor to metal. This type of junction is an example of a *Schottky barrier*.

5.4.2. *Behaviour in the light*

Now suppose that the semiconductor is illuminated with photons of energy greater than E_g. The space charge layer will cause electron-hole pairs generated in the semiconductor to be separated so that the electrons accumulate in the semiconductor and the holes in the metal (*i.e.*, electrons removed from the metal). This applies to pairs generated far from the field bearing

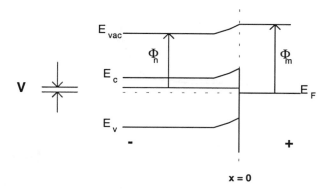

Fig. 5.5. Band profile of the semiconductor–metal junction under illumination at open circuit. The accumulation of photogenerated electrons in the n-type semiconductor raises the electron Fermi level and generates a photovoltage, V.

region through the diffusive current resulting from preferential hole extraction at the interface. The semiconductor will become negatively charged and the potential difference across the junction will be reduced. The electron quasi Fermi level far from the junction will be higher than it was in the dark, and higher than the Fermi level in the metal, as shown in Fig. 5.5. The light has caused the Fermi level to split. It has created a *photovoltage*, V, equal to the difference in the Fermi levels of semiconductor and metal far from the junction. This ability to sustain a difference in quasi Fermi levels under illumination is the key requirement for photovoltaic energy conversion.

5.4.3. *Behaviour in the dark*

The presence of the barrier also governs the current–voltage characteristic in the dark. Conduction in n type semiconductors is normally by electron flow, and the interface presents a barrier to electron flow. At equilibrium, a small current due to the thermal activation of electrons over the barrier is balanced by a small 'leakage' current due to the drift of holes from semiconductor to metal. When a forward bias is applied (the semiconductor is held at a more negative bias than the metal) the barrier height is reduced and electrons pass more easily over the barrier from semiconductor to metal. This forward current increases approximately exponentially as the barrier height is reduced. In reverse bias, the barrier height is increased, suppressing the activated current, and the only contribution is the small

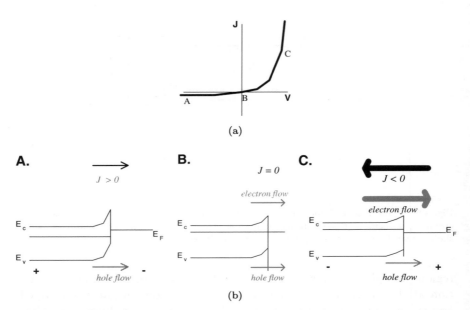

Fig. 5.6.   (a) Schematic current–voltage characteristic of a Schottky barrier junction in the dark. A, B and C mark points on the curve where the device is at reverse bias, equilibrium and forward bias; (b) A: band profile of an n-type semiconductor–metal Schottky barrier at reverse bias. The only current is due to minority carrier (hole) drift across the depleted barrier region. B: band profile at equilibrium. The currents due to electron diffusion and hole drift cancel out. C: band profile at forward bias. The current due to electron diffusion is greatly increased as the barrier height is reduced, and the net current changes sign.

leakage current in the reverse direction, which is limited by the low density of mobile holes. So, the junction passes current preferentially in the forward direction, and exhibits 'rectifying' characteristics (Fig. 5.5). (This was the special property of the early metal–semiconductor junctions mentioned in Chapter 1, to which photovoltaic behaviour was attributed.) This asymmetric current–voltage behaviour in the dark is a consequence of the charge separating junction and is a feature of most photovoltaic devices. The greater the difference in work functions, the stronger the band bending and the greater the asymmetry between the forward and reverse currents.

n and p type metal–semiconductor junction

A completely analogous situation applies to a p type semiconductor in contact with a metal of lower work function, *i.e.*, $\Phi_m < \Phi_p$. In this case the

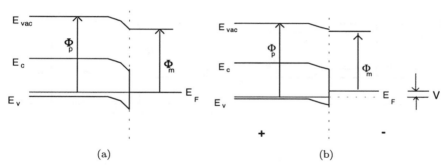

Fig. 5.7. Band profile of the *p*-type semiconductor–metal junction (a) at equilibrium and (b) under illumination at open circuit.

semiconductor bands bend *down* towards the interface (Fig. 5.7), presenting a barrier to majority carriers which are now *holes*. The depleted layer of the semiconductor is *negatively* charged due to the space charge of the ionised acceptor impurities. The forward current is provided by hole activation over the barrier, and the reverse current by electron leakage. Under illumination, electrons are driven into the metal and holes into the semiconductor, and the semiconductor develops a photovoltage which is positive relative to the metal. In the dark, the junction exhibits the same type of asymmetric current voltage behaviour as the *n* type Schottky barrier, but with opposite polarity.

5.4.4. *Ohmic contacts*

The situation is different if we have an *n* type-metal junction such that $\Phi_m < \Phi_n$, or a *p* type metal junction with $\Phi_m > \Phi_p$. Now, when the materials are brought into contact, the semiconductor bands bend in such a way that they encourage the transport of majority carriers across the junction, and inhibit only the flow of minority carriers (Fig. 5.8). Majority carriers must accumulate near the interface to establish the necessary potential difference, and the junction region is rich with carriers. This means it can pass current easily in either direction, and so we have a low resistance contact for majority carriers, usually called an *Ohmic* contact. Under illumination, the charges separated at the junction pass back across the junction relatively easily, so that the resultant photovoltage at the terminals is negligible. The mechanism which gives rise to the photovoltage in a barrier junction — the selective removal of minority carriers — is absent.

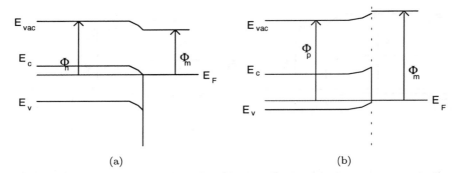

(a)   (b)

Fig. 5.8. Ohmic metal–semiconductor contacts for (a) an n-type semiconductor and (b) a p-type semiconductor. In each case the difference in work functions is supplied by the build up of majority carriers in an accumulation layer near the interface. An accumulation layer is generally narrow compared to a depletion layer because of the higher density of charges and stronger electric field.

We can draw the following conclusions from the above discussion:

- a charge separating field is established at the interface between two materials of different work function
- the junction will develop a photovoltage provided that it presents a barrier to majority carrier currents
- the photovoltage is related to the difference in work functions

5.4.5. *Limitations of the Schottky barrier junction*

The metal–semiconductor barrier is a simple way to prepare a photovoltaic junction. However, it does not produce the highest photovoltages; (i) When the barrier height is larger than $\sim E_{\mathrm{g}}/2$, the minority carriers outnumber the majority carriers in the region close to the interface and an *inversion layer* is formed. In these conditions the the junction becomes carrier rich and cannot sustain a photovoltage. This limits the useful barrier height, and hence limits the photovoltage; (ii) In the case of highly doped semiconductors, very thin barrier layers may result which allow tunnelling of majority carriers through the junction, reducing the effectiveness of the barrier; (iii) The material interface between the semiconductor and the metal leads to interface states which may trap charge and limit the photovoltage, as discussed below.

The problems of the Schottky barrier can be avoided with semiconductor–semiconductor junctions.

5.5.   Semiconductor–Semiconductor Junctions

5.5.1.   *p–n junction*

The *p–n* junction is the classical model of a solar cell. This type of junction is created by doping different regions of the same semiconductor differently, so we have an interface between *p* type and *n* type layers of the same material. Since the work function of the *p* type material is larger than the *n* type, the electrostatic potential must be smaller on the *n* side than the *p*, and an electric field is established at the junction (see Fig. 5.9) which drives photogenerated electrons towards the *n* side and holes towards the *p*. The junction region is depleted of both electrons and holes, and always presents a barrier to majority carriers, and a low resistance path to minority carriers. It drives the collection of minority carriers which are photogenerated throughout the *p* and *n* layers, and reach the junction by diffusion.

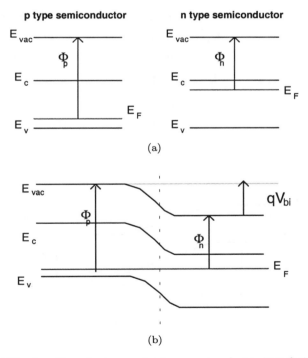

Fig. 5.9.   (a) Band profiles of *p*-type and *n*-type semiconductor in isolation. (b) Band profile of the *p–n* junction in equilibrium.

The p–n junction has the advantages over the Schottky barrier that (i) there is no need for a metallurgical interface with its associated defects. The junction can be created by continuously varying the doping of a single crystal wafer during or after growth; (ii) A large built in bias can be established without populating the junction region (no inversion layer).

The p–n junction is the most widely used structure for solar cells and will be analysed in detail in Chapter 6.

5.5.2.   *p–i–n junction*

A variation on the p–n theme is the p–i–n junction. This is a p–n junction where a layer of semiconductor between p and n has been left undoped, or intrinsic (i). The same built-in bias is established as with the p–n junction with the same doping levels, but the electric field extends over a wider region, as shown in Fig. 5.10. This design is preferred in materials where the

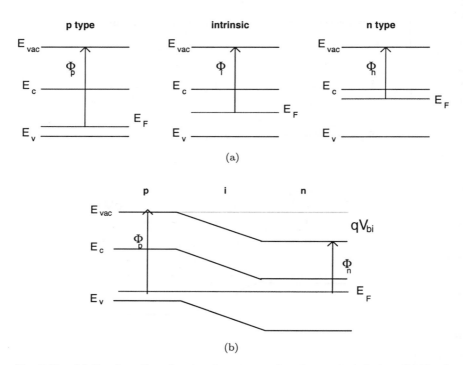

Fig. 5.10.   (a) Band profiles of p, i and n-type semiconductors in isolation. (b) Band profile of the p–i–n junction in equilibrium.

minority carrier diffusion lengths are short, and carriers photogenerated in p or n layers are unlikely to contribute to the photocurrent. Carriers which are photogenerated in the i region are driven towards the contacts by the electric field and survive for a greater distance than in doped material. Lifetimes of carriers in the i region are usually extended relative to those in the doped regions. Some disadvantages of the p–i–n design are that (i) the i region has poorer conductivity than doped layers, and may introduce series resistance; (ii) the likelihood of recombination within the i layer at forward bias conditions, where electron and hole populations becomes similar; and (iii) charged impurities may cause the electric field to fall to zero within the intrinsic region. p–i–n junctions in amorphous silicon will be discussed in Chapter 8.

5.5.3. *p–n heterojunction*

p–n and p–i–n junctions can also be prepared as *heterojunctions* using two different materials of different band gap. This might be a design feature to improve carrier collection (discussed in Chapter 7) or a necessity because of the doping properties of available materials. At the junction there will be a discontinuity in the conduction and valence band edges due to the change in the band gap. This potential step introduces different effective fields for electrons and holes which usually assist the electrostatic field for one carrier and oppose it for the other. In the example in Fig. 5.11, the heterojunction enhances the field driving electrons towards the n side but opposes the field for holes by introducing a barrier in the valence band. Such barriers are common — though not unavoidable — in p–n heterojunctions, and can lead to enhanced recombination in the junction region. The exact band line-up will depend upon the difference in the electron affinities and the work functions.

5.6. Electrochemical Junction

A surface barrier junction similar to the Schottky barrier can be established at the interface between a semiconductor and a liquid electrolyte. An electrolyte which contains a *redox couple* is capable of transporting charge to and from the semiconductor surface. The redox couple contains two ionic species of different state of charge, or oxidation state. These may be ions of the same material in different states of charge, for instance Fe^{2+} and Fe^{3+}. The one with a more positive charge is the oxidised species, while

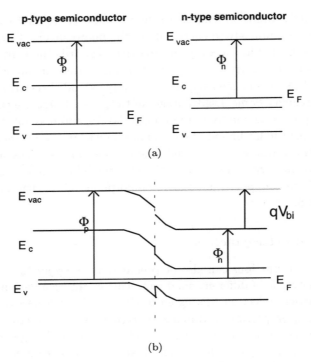

Fig. 5.11. (a) Band profiles of p and n-type semiconductors of different band gap in isolation; (b) Band profile of the p–n heterojunction in equilibrium.

the more negative one is the reduced species. The species are capable of exchanging charge with each other by transferring an electron. The relative concentrations are fixed on preparation (like doping a semiconductor) and the overall charge is balanced with counter ions.

Like a metal or semiconductor, the redox couple has a Fermi level, or chemical potential, which is constant at equilibrium. To predict the nature of the junction, we need to know the effective work function of the electrolyte, Φ_e, which is the difference between the chemical potential and the vacuum level. Redox potentials are usually given relative to a standard reference level, such as the redox potential of hydrogen. If the work function of the reference is known, then the work function of the electrolyte can be defined from the redox potential [Fonash, 1980]

$$\Phi_e = E^0_{\text{redox}} + \Phi_{\text{ref}} - k_B T \ln\left(\frac{[\text{Red}]}{[\text{Ox}]}\right) \qquad (5.7)$$

where E^0_{redox} is the standard redox potential of the electrolyte and Φ_{ref} is the work function of the standard. The final term shows how the work function depends upon the relative concentrations of reduced species, [Red], and oxidised species, [Ox]. (The standard redox potential is defined for equal concentrations.)

The type of junction which is formed at a perfect interface depends upon the difference in the work functions of the semiconductor and the electrolyte. Just as with the semiconductor–metal junction, a surface barrier will be formed at the interface between an n type semiconductor with work function Φ_n and an electrolyte with $\Phi_e > \Phi_n$. This is illustrated in Fig. 5.12(b). Upon contact, electrons flow from semiconductor into electrolyte until the Fermi levels equalise, establishing a positive space charge layer in the semiconductor and an electric field at the interface which drives

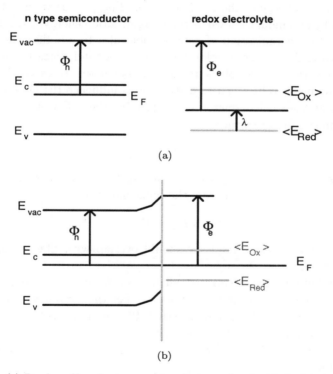

Fig. 5.12. (a) Band profiles of n-type semiconductor and redox electrolyte in isolation; (b) Band profile of the semiconductor–electrolyte heterojunction in equilibrium. A depletion layer is formed at the semiconductor surface, as in a Schottky barrier. λ represents the reorganisation energy.

charge separation. A balancing charged layer is formed in the electrolyte. This charged layer is usually taken up within the first few monolayers of solvent molecules, in a region called the Helmholtz layer which may sustain large electric field. The ionic concentrations may continue to be disturbed from equilibrium through a region beyond the Helmholtz layer called the diffuse layer. Because the permittivites of electrolytes, like metals, are small, the difference in work functions is taken up almost entirely by the semiconductor, *i.e.*, the potential drop within the Helmholtz layer of the solution, V_H, is small and

$$\frac{1}{q}\Delta\Phi = V_{bi} + V_H \approx V_{bi}\,.$$

The energy levels of the reduced and oxidised species in the electrolyte are slightly different. The mean energy of the oxidised state, $\langle E_{Ox}\rangle$ is higher and the mean energy of the reduced state, $\langle E_{Red}\rangle$ is lower than the redox potential by an amount known as the reorganisation energy. This difference is due to the reorgnisation of solvent molecules to stabilise the electron on the reduced ion, like a polaronic effect in a solid. Because of thermal fluctuations, the energies of the reduced and oxidised states are distributed around the mean values. These distributions influence the charge transfer processes at the semiconductor–electrolyte interface. For electron transfer to the semiconductor surface,

$$\text{Red} \rightarrow \text{Ox} + e_{sc}^-\,,$$

the levels of the reduced species should overlap with vacant levels in the semiconductor. For transfer from the semiconductor,

$$\text{Ox} + e_{sc}^- \rightarrow \text{Red}\,,$$

the levels of the oxidised species should overlap with filled levels in the semiconductor.

Under illumination, electrons will be transferred to or from the semiconductor according to the type of doping. For the n-type semiconductor–electrolyte junction illustrated in Fig. 5.12, electrons will be transferred to the semiconductor surface, resulting in the semiconductor gaining a negative charge and the electrolyte a positive charge, so providing a photovoltage.

The current density in the electrolyte is due to the net flux of reduced species (in this example) towards and oxidised species away from the semiconductor surface. This provides a current directed from semiconductor to electrolyte. For current to flows through the external circuit, the oxidised species must be able to recover an electron from a counter electrode and regenerate the reduced form.

The electrochemical junction is a poor man's Schottky barrier. It has the advantage that the field is established spontaneously upon wetting the semiconductor surface, but the disadvantage that, in many materials systems, the semiconductor surface is prone to react chemically with the electrolyte under illumination. These problems can be avoided with sensitised electrochemical junctions [Grätzel, 2001].

5.7. Junctions in Organic Materials

Junctions in molecular photovoltaic materials require different considerations. In a molecular semiconductor light generates excitons which are relatively strongly bound, depending on the strength of the intra-molecular forces compared to those binding the molecules together. In some crystalline organic solids, intermolecular forces are strong and carriers may be considered to occupy bands much like inorganic crystals. In such materials, excitons may be split spontaneously and devices can be designed using similar principles as for inorganic metal–semiconductor junctions. In other materials, such as amorphous organic solids or polymers, intra-molecular forces dominate and the excitons are very tightly bound. In such cases the electrostatic fields available from the difference in work functions of the junction materials is not usually sufficient to split the exciton. Instead, the excitons drift, and only split when they approach the junction with a contact material of different work function. Charge separation thus occurs only at the junction. The problem is that a tightly bound exciton is likely to recombine before it reaches the junction. In typical molecular materials the exciton diffusion length is a few tens of nanometres. This means that for a Schottky barrier type structure, only the 10 nm of material closest to the junction can contribute to the photocurrent. Hundreds of nm of the material will be needed for a good optical depth.

A solution to this problem is to use a distributed interface, where two materials of different electronic structure are combined into a blend. The materials should be chosen so that their energy levels line up to encourage excitons to split at the interface, as shown in Fig. 5.13. The difference in electron affinities leads to strong local electric fields. Then when an exciton diffuses to the interface it is likely to be split and the resulting charge states (which are polarons in molecular materials) are free to diffuse towards the

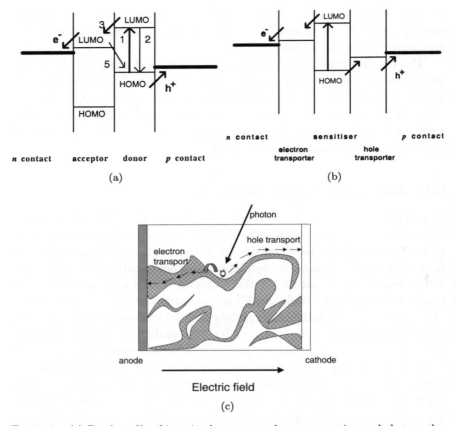

Fig. 5.13.   (a) Band profile of junction between an electron accepting and electron do-
nating molecular semiconductor under illumination. Photon absorption (1) is followed
by recombination (2) or exciton dissociation (3). Separated charges may recombine (5).
(b) Band profile for a donor–acceptor junction with a sensitising molecule to absorb the
light; (c) Schematic of charge separation and transport through a donor–acceptor blend.

collecting contacts. Charge separation will be efficient if the materials are
blended on a scale which is similar to the exciton diffusion length, *i.e.*, on
the nanometre scale. By using a distributed junction, charge separation
can be achieved throughout a layer of high optical depth. As in all photovoltaic structures, it is necessary to use asymmetric contacts to the blended
materials, so as to collect electrons and holes selectively.

    A similar principle is exploited in dye sensitised devices, where a molec-
ular sensitiser is anchored at the interface between an electron transporting
semiconductor and a hole transporting medium. In this case the junction is

the sensitiser-semiconductor interface where the difference in electron affinities splits the exciton. Nanocrystalline porous semiconductor substrates allow blending of the materials on a 1–100 nm scale. The physics of organic photovoltaic devices is described by Halls and Friend [Halls, 2001], and the principles of dye sensitised junctions by Grätzel [Grätzel, 2001; Hagfeldt and Grätzel, 2000].

5.8. Surface and Interface States

In discussing the various types of interface (metal–semiconductor, semiconductor heterojunction, semiconductor–electrolyte) we have supposed the semiconductor interfaces are perfect. In fact, solid surfaces are likely to contain defects and impurities, both intrinsic defects due to the interruption of the crystal structure at the surface, and extrinsic defects due, *e.g.* to adsorbed impurity atoms. These defects tend to introduce extra electronic states with energies in the band gap. These 'surface' or 'interface' states are (spatially) localised near to the interface and are capable of trapping charge, and so they will influence the potential distribution across the interface through Eq. 5.6. When interface states are present, the potential difference necessary to match work functions across a junction will be distributed differently.

5.8.1. *Surface states on free surfaces*

Intra-band gap states near to the valence band tend to trap electrons (acceptor states) while states near to the conduction band tend to trap holes (donor states). In general, interface states may be distributed in energy throughout the band gap region, and whether they act like acceptors or donors depends upon the occupancy of the states and the Fermi level of the semiconductor. It is useful to define a neutrality level ϕ_0 as the level up to which the states are filled when the surface (or interface) is neutral. This is like a Fermi level for the surface in isolation. When the surface adjoins a layer of semiconductor, ϕ_0 is in general at a different energy than the semiconductor Fermi level and so charge must be exchanged to bring the surface and semiconductor into equilibrium. If $\phi_0 < E_F$ the surface is acceptor like and traps electrons, while if $\phi_0 > E_F$ the surface is donor like and traps holes. Figure 5.14 shows how a depletion region is established at the surface of an n type semiconductor for which $\phi_0 < E_F$. Upon contact, electrons flow from semiconductor into surface states, leaving a positively

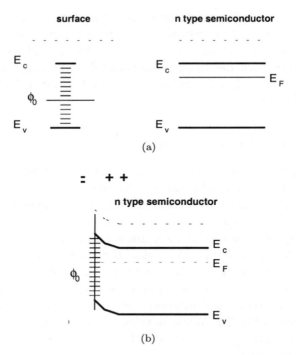

Fig. 5.14.   (a) Band profiles of surface and n-type semiconductors in isolation; (b) Band profile of the surface in equilibrium.

charged depleted layer behind until the potential barrier established, V_{bi}, is big enough to prevent further flow. The surface charge accumulated is negative and is given by

$$-Q_s = -q \int_{\phi_0}^{E_c - V_n - V_{bi}} g_s(E) dE \qquad (5.8)$$

where V_n is the donor ionisation energy, $g_s(E)$ is the density of surface states and V_{bi} is determined by simultaneously solving Poisson's equation. This must be matched by a positive charge $+Q_s$ inside the semiconductor.

A negative depletion layer is similarly established near the surface of a p-type semiconductor with $\phi_0 < E_F$. See Bardeen [Bardeen, 1947], for an early discussion of the effects of surface states.

If instead we had a surface with $\phi_0 > E_F$ on an n-type semiconductor, or $\phi_0 < E_F$ on a p-type semiconductor, then the semiconductor would accumulate carriers to compensate the surface charge, but the charged layer would extend much less far into the semiconductor and the band bending would be smaller.

5.8.2. *Effect of interface states on junctions*

From the above we see that surface or interface states can introduce a potential difference, and an electric field into an otherwise neutral semi-conductor structure. At a junction, interface states may trap charge and influence the potential and electric field distribution. Although interface states cannot alter the net potential difference across the junction (which must always equal the difference in work functions), they influence the way in which that potential is divided between the two materials.

Interface states at a p–n junction

Suppose we have an acceptor like interface at a *p–n* junction. When the junction is formed, some electrons from the *n*-type layer become trapped at the interface, so that a smaller negative space charge needs to be developed in the *p* side to compensate the positive space charge on the *n* side. The consequence is that more of the potential difference is dropped on the *n* side, and the electric field, which is discontinuous at the charged interface, is now higher on the *n* side. To support the larger potential difference on the *n* side an *inversion layer* of mobile holes (minority carriers) may accumulate beside the interface.

Figure 5.15 shows how the band profile may change. Notice that the effect of the interface layer is to pull E_c closer to E_F at the interface.

We should identify two extreme situations:

- If the density of interface states is large enough, then the Fermi level at the interface will be pinned at ϕ_0, irrespective of the work functions of the materials on either side. The high density of trapped charge at the interface screens the two materials from each other.
- If, in addition, ϕ_0 is low enough (comparable with the acceptor ionisation energy V_p), then the entire potential difference can be dropped on the *n* side.

Interface states at a metal–semiconductor junction

An interface layer immediately next to the metal at a metal–semiconductor junction has no effect on the electrostatics of the junction because the

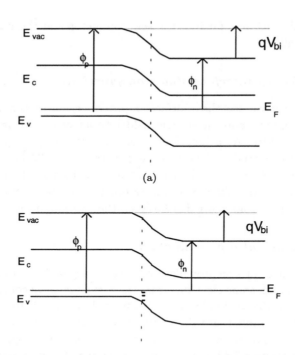

Fig. 5.15. (a) Band profiles of a p–n junction in equilibrium (a) in the absence of interface states (b) with acceptor like interface states, which change the division of the built-in bias between the two semiconductors.

interface charge will be screened by the infinitely mobile charge of the metal. If, however, the interface layer is separated from the metal by an insulating layer then charge can be stored at either side of this insulator, like a parallel plate capacitor, causing a potential drop, V_{int}. This is a realistic situation since a thin layer of oxide may form between the metal and the semiconductor. V_{int} contributes to the change in work functions across the junction through

$$\frac{1}{q}\Delta\phi_{w} = V_{bi} + V_{int} \tag{5.9}$$

and must be given by

$$V_{int} = \varepsilon Q d$$

where Q here represents the total charge developed in the semiconductor, d the thickness of the interface layer and ε is the dielectric permittivity of the insulating layer. Note that the presence of the insulating layer will always mean that the built-in bias in the semiconductor, V_{bi}, is less than $\frac{1}{q}\Delta\Phi_w$. The role of interface states is to influence the sharing of the potential difference between the semiconductor and the insulating layer. With an n-type semiconductor, acceptor like states tend to reduce the electric field at the interface so that less of the potential difference is dropped in the insulator and more in the n-type material, just as in the case of the heterojunction (*i.e.*, V_{bi} is increased relative to the case with no interface states). Donor like states increase the field at the interface and reduce V_{bi}. For an extremely high density of interface states, E_F will be pinned at ϕ_0, the interface charge will screen the metal from the semiconductor and V_{bi} will be independent of the work function of the metal. This is called the Bardeen limit for metal semiconductor junctions. In the opposite limit, of a perfect junction with no interface states, any change in the work function of the metal causes an equal change in V_{bi}.

Interface states at a semiconductor–electrolyte junction

Just as at the semiconductor–metal junction, interface states at a semiconductor–electrolyte junction can accommodate part of the difference in work functions between the semiconductor and redox couple. Including the interface state charge we have

$$\Delta\phi_w = V_{bi} + V_{int} + V_H \,.$$

Again, for a very high density of interface states, E_F is pinned at ϕ_0 at the interface and V_{bi} is independent of the redox couple.

5.9. Summary

For a light absorbing device to work as a solar cell, some kind of asymmetry needs to be built in to the system which drives electrons and holes in different directions. This asymmetry may be introduced through spatial variations in the electronic properties. Spatial variation in electron affinity, band gap, work function or density of states can drive charge separation. Variations in work function can generate large electric fields and are most commonly used.

At the interface between two different electronic materials, a potential step equal to the difference in work function is established by exchange of charge carriers. The potential is taken up in the semiconductor(s) and is represented by bending the bands. The junction is effective for charge separation if the semiconductor is depleted of carriers and blocks the flow of majority carriers across the junction. If the semiconductor is enriched with charge the junction is Ohmic. A blocking contact can be formed between a metal and semiconductor (a Schottky barrier), between an n and p-type semiconductor (a p–n junction) or between a semiconductor and an electrolyte. When the device is illuminated, electrons and holes are separated by the field at the junction, and accumulate on opposite sides. The photovoltage is determined by the maximum charge separation and is related to the height of the potential step and the rectifying action of the junction. At a Schottky barrier the photovoltage is limited by the formation of an inversion layer and by tunnelling. p–n junctions perform better. Defects at the interface introduce states in the semiconductor band gap which can trap charges and influence the potential distribution at the junction. A high density of interface states reduces the photovoltage and degrades the photovoltaic performance of the device.

References

J. Bardeen, "Surface states", *Phys. Rev.* **71**, 717 (1947).

S.J. Fonash, *Solar Cell Device Physics* (New York: Academic, 1980).

M. Grätzel, "Photoelectrochemical cells", *Nature* **414**, 338–344 (2001).

A. Hagfeldt and M. Grätzel, "Molecular photovoltaics", *Acc. Chem. Res.* **33**, 269–277 (2000).

J.J.M. Halls and R.H. Friend, "Organic photovoltaic devices" in *Clean Electricity from Photovoltaics*, eds. M.D. Archer and R.D. Hill (Imperial College Press, 2001).

Chapter 6

Analysis of the p–n Junction

6.1. Introduction

The p–n homojunction is the most widely used device structure in photovoltaics. Selective doping of the different sides of a semiconductor wafer p type and n type leads to a potential barrier between the regions. This junction acts as a selective barrier to charge carrier flow, so that there is a low resistance path for electrons to the n contact and for holes to the p contact, thus providing the asymmetry in resistance which is necessary for photovoltaic conversion. By control of the doping levels, large potential barriers can be established which make it possible to generate large photovoltages. It has the practical advantage that the junction can be prepared from a single crystal semiconductor wafer without the need for a metallurgical interface, and the associated problems of interface states.

The photovoltaic function of a p–n junction can be analysed using the theoretical apparatus introduced in Chapters 3 and 4. In general, the problem consists of solving a set of coupled differential equations for the electron density, hole density and the electrostatic potential, given specified forms for the photogeneration, the recombination and the electron and hole currents. It is in general a complex problem but, with the benefit of two approximations, analytic solutions for the $J(V)$ characteristic under different conditions can be found. These two approximations are the depletion approximation, and the approximation that recombination in the doped material is linear. This allows the differential equations to be uncoupled and linearised. In the next sections we first outline the approach to solving the p–n junction, before presenting the solutions and discussing some of the implications for junction design.

6.2.   The *p–n* Junction

6.2.1.   *Formation of p–n junction*

As explained in Chapter 5, a *p–n* junction is established when a layer of *p*-type semiconductor and a layer of *n*-type material are brought together. Carriers diffuse across the junction leaving behind a layer of fixed charge, due to the now ionised impurity atoms, on either side. This space charge sets up an electrostatic field which opposes further diffusion across the junction. Equilibrium is established when diffusion of majority carriers across the junction is balanced by the drift of minority carriers back across the junction in the built in electrostatic field. At this point the Fermi levels of the *p* and *n* type layers are equal, the difference in the work functions is taken up by a step in the conduction and valence band edges, called the built-in bias, and the junction region is depleted of carriers.

The built-in bias in equilibrium, V_{bi}, is determined by the difference in work functions of the *n* and *p* type materials, Φ_n and Φ_p. The difference in work functions is equal to the difference in the shift of the Fermi levels from the intrinsic potential energy of the semiconductor, since E_i is parallel to E_{vac}. Hence

$$V_{bi} = \frac{1}{q}(\Phi_n - \Phi_p) = \frac{1}{q}((E_i - E_F)|_{p \text{ side}} - (E_i - E_F)|_{n \text{ side}}). \quad (6.1)$$

Far from the junction, $n = N_d$ on the *n* side and $p = N_a$ on the *p* side. The shifts $E_i - E_F$ can then be expressed in terms of the doping levels using Eqs. 3.37 and 3.38 for *n* and *p*. This gives for V_{bi},

$$V_{bi} = \frac{k_B T}{q} \ln \left(\frac{N_d N_a}{n_i^2} \right) \quad (6.2)$$

relating V_{bi} to the doping levels. Figure 6.1 illustrates the space charge and band bending in a *p–n* junction at equilibrium.

If an external electrical bias is applied, it raises the Fermi level on one side with respect to the other and the potential dropped across the junction, V_j becomes

$$V_j = V_{bi} - V \quad (6.3)$$

where V is the bias applied to the *p* side. A negative applied bias withdraws carriers from the junction region, increasing the depletion while a positive applied bias injects carriers and reduces the depletion. A positive

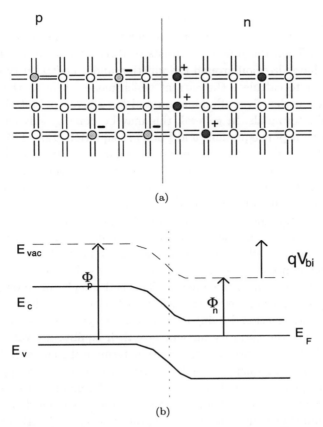

(a)

(b)

Fig. 6.1. (a) Schematic of crystal structure close to a *p–n* junction. Close to the junction free carriers are removed, so that acceptor impurities on the *p* side (grey circles) become negatively charged, and donor impurities on the *n* side become positively charged; (b) Energy band profile across the junction in equilibrium.

bias of V_{bi} will completely cancel the potential step and remove the asymmetry which drives the photovoltaic effect. Illumination of the semiconductor causes photogenerated holes to build up in the *p* side and electrons on the *n* side, shifting the Fermi levels in the same sense as a positive applied bias.

6.2.2. Outline of approach

In the analysis of the *p–n* junction which follows, we assume that we have high quality, crystalline semiconductor layers, that the junction is free of interface states. We assume that the built in bias can be taken up completely within the n and p layers. This is true for typical doping levels and layer thicknesses. Then the junction consists of three regions: a neutral p type region, a charged region around the junction and a neutral n type region.

We want to calculate the current which passes through the device in steady state under given illumination and for a given potential difference between the terminals. In general, this may be done by solving the semiconductor transport equations and Poisson's equation (Eqs. 4.87–4.89) self consistently together with appropriate boundary conditions at the p and n terminals and appropriate forms for the photogeneration and recombination rates and the electron and hole currents. The problem is, in general, complicated because of the coupling of electron and hole densities through Poisson's equation and non-linear recombination terms.

The problem is simplified by making two approximations. First, we assume that the charged region around the junction contains no free carriers so that the potential step is taken up completely by the fixed space charge of the doped materials near the junction. This is the depletion approximation. In this approximation, the electric field vanishes at a fixed distance from either side of the junction, leaving neutral p and n type regions. Within these neutral regions, the majority carriers have their equilibrium value and only variations in the minority carrier density determine the current. Across the depleted region the quasi Fermi levels are separated by the applied bias, qV_j and this provides a boundary condition on the minority carrier density at the edge of each neutral region, so the minority carrier density can be solved within that region alone. Finally, the absence of electric field means that the minority carrier currents are driven by diffusion only and the current equations are simplified. Thus the depletion approximation allows the solutions in the neutral p and n regions to be decoupled.

The second approximation is that the recombination rates in the neutral regions are linear in the minority carrier density. This is reasonable, according to Eqs. 4.80–4.81, since the majority carrier densities are large and constant. This allows analytic solutions to be found. This linearisation decouples the effect of bias from the effect of illumination, so that the solutions under bias and under light may be added to give the solution under both light and bias. This is sometimes called the superposition approximation.

In the following, we will first establish the width of the depleted region, or space charge region (SCR). Then we set up the equations for the minority carrier densities in the neutral regions, and find the general solution for the minority carrier currents under illumination and applied bias. We include terms for the dark and photo-currents within the depletion region. Before the depletion region current is included, the current voltage characteristic has the ideal diode form (Eq. 1.5) and the superposition approximation (Eq. 1.4) is obeyed.

We look at solutions for special cases to show how the photocurrent is influenced by the minority carrier transport parameters and how the dark current is influenced by the recombination mechanism. Series resistance can be accommodated by adjusting the junction potential to allow for potential drops in the doped regions, and the effects of temperature and light intensity are studied.

6.3. Depletion Approximation

We define our junction as a layer of p type material of doping N_a for $x < 0$ which adjoins a layer of n type material of doping N_d for $x > 0$, with a perfect interface in the plane $x = 0$. The p layer has thickness x_p and the n layer has thickness x_n. The junction is completely free of majority carriers for a depth w_p into the p layer and w_n into the n layer, as shown in Fig. 6.2. Then the depleted layers in the two materials are charged by the ionised impurity atoms. The regions beyond these boundaries are completely neutral.

Note that an n–p junction can be treated exactly the same way: only the direction of x changes.

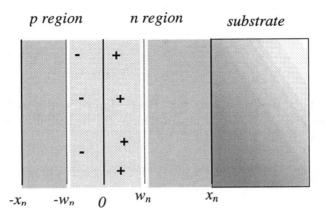

Fig. 6.2. Schematic layer structure of p–n junction, showing charged region near junction.

6.3.1.   *Calculation of depletion width*

According to Poisson's equation the electrostatic potential ϕ must obey

$$\frac{d^2\phi}{dx^2} = \frac{q}{\varepsilon_s}N_a \qquad \text{for } x < 0 \tag{6.4}$$

and

$$\frac{d^2\phi}{dx^2} = -\frac{q}{\varepsilon_s}N_d \qquad \text{for } x > 0 \tag{6.5}$$

where ε_s is the permittivity of the semiconductor and ϕ is related to the intrinsic potential energy E_i through $E_i = -q\phi + C$ where C is any constant and can be set to 0. Since ϕ can vary only within the space charge region it must change by V_{bi} across the width of that region. This gives us the boundary conditions

$$\phi = 0 \qquad \text{at } x = -w_p$$

and

$$\phi = V_{bi} \qquad \text{at } x = w_n.$$

Integrating Poisson's equation then yields expressions for the electric field, $F = -\frac{d\phi}{dx}$,

$$-\frac{d\phi}{dx} = -\frac{qN_a}{\varepsilon_s}(x + w_p) \qquad \text{for } -w_n < x < 0 \tag{6.6}$$

and

$$-\frac{d\phi}{dx} = \frac{qN_d}{\varepsilon_s}(x - w_n) \qquad \text{for } 0 < x < w_p \tag{6.7}$$

using the boundary conditions that the field must vanish at the edges of the depleted layers. A further integration, using the boundary condition on ϕ yields

$$\phi = \frac{qN_a}{2\varepsilon_s}(x + w_p)^2 \qquad \text{for } -w_n < x < 0 \tag{6.8}$$

and

$$\phi = -\frac{qN_a}{2\varepsilon_s}(x - w_n)^2 + V_{bi} \qquad \text{for } 0 < x < w_p. \tag{6.9}$$

Requiring that ϕ and F are continuous at $x = 0$ we obtain the individual depleted widths

$$w_\mathrm{p} = \frac{1}{N_\mathrm{a}} \sqrt{\frac{2\varepsilon_\mathrm{s} V_\mathrm{bi}}{q(\frac{1}{N_\mathrm{a}} + \frac{1}{N_\mathrm{d}})}}, \tag{6.10}$$

$$w_\mathrm{n} = \frac{1}{N_\mathrm{d}} \sqrt{\frac{2\varepsilon_\mathrm{s} V_\mathrm{bi}}{q(\frac{1}{N_\mathrm{a}} + \frac{1}{N_\mathrm{d}})}}, \tag{6.11}$$

and the total width of the space charge region, w_scr

$$w_\mathrm{scr} = w_\mathrm{p} + w_\mathrm{n} = \sqrt{\frac{2\varepsilon_\mathrm{s}}{q} \left(\frac{1}{N_\mathrm{a}} + \frac{1}{N_\mathrm{d}} \right) V_\mathrm{bi}}. \tag{6.12a}$$

Notice that the depleted width increases as either the p doping or the n doping is reduced. This means that it is not feasible to have a p–n junction with both a wide depleted region, which aids carrier collection, and high doping levels, which aid cell voltage. This is one of several compromises in cell design.

Notice also that the division of V_bi between the two layers depends upon the relative doping: more of V_bi is dropped in the layer which has the lower doping and the wider depleted width. In most crystalline solar cells it is usual to dope the top layer (known in electronics terminology as the *emitter*) heavily but the lower layer (known as the *base*) lightly, so that almost all of the space charge region lies within the base. The reasons for this will be explained in Chapter 7. Then, for a p on n structure,

$$w_\mathrm{scr} \approx w_\mathrm{n} = \sqrt{\frac{2\varepsilon_\mathrm{s}}{q N_\mathrm{d}} V_\mathrm{bi}}. \tag{6.12b}$$

The potential and electric field profile for a typical junction are illustrated in Fig. 6.3.

If a forward bias V is applied to the p side relative to the n side, then the bias across the depleted region is reduced to $V_\mathrm{j} = (V_\mathrm{bi} - V)$. The boundary condition at w_n becomes $\phi(w_\mathrm{n}) = V_\mathrm{j}$, and w_n and w_p will be reduced in proportion to $\sqrt{V_\mathrm{bi} - V}$. Similarly, if the junction is reverse biased, the potential drop increases and the depleted layers become thicker. The applied bias shifts the relative position of the quasi Fermi levels of the *majority* carriers in the neutral regions, so that

$$V = \frac{1}{q}(E_{\mathrm{F_n}}(w_\mathrm{n}) - E_{\mathrm{F_p}}(-w_\mathrm{p})). \tag{6.13}$$

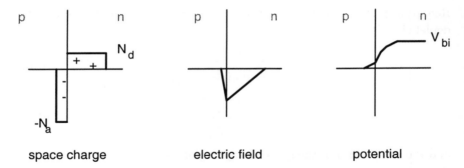

space charge   electric field   potential

Fig. 6.3. Doping profile, electric field and potential variation across a *p–n* junction. The electric field varies linearly and the electrostatic potential quadratically across each depleted region. In this example, the *p* layer is doped more heavily than the *n* and so the depletion width is larger in the *n*.

Because all potential is dropped within the depleted layer, the quasi Fermi levels of majority carriers remain constant throughout the neutral regions. A further requirement of depletion is that the difference in electron and hole quasi Fermi levels must remain constant and equal to V across the depleted layer, *i.e.*,

$$pn = n_i^2 e^{qV/k_B T} \qquad \text{for } -w_n < x < w_p .$$

This has the consequence that E_{F_n} and E_{F_p} are both constant across the depleted region.

6.4.   Calculation of Carrier and Current Densities

6.4.1.   *Currents and carrier densities in the neutral regions*

To calculate the carrier densities we will use the semiconductor transport equations derived in Chapter 4. These simplify for each of the three regions of the *p–n* junction, which we will consider separately. We will consider a junction illuminated by a flux density $b_s(E)$ of photons of energy E, and subject to an applied bias V. We will use $j_n(E, x)$, $j_p(E, x)$ to denote the solutions for the electron and hole current densities, respectively, at depth x under monochromatic illumination. The current densities for the panchromatic spectrum are easily constructed, through

$$J_n(x) = \int j_n(E, x) dE \qquad (6.14a)$$

$$J_p(x) = \int j_p(E, x) dE \qquad (6.14b)$$

provided that the differential equations governing n and p are linear.

In the electrically neutral p and n layers, the electric field is zero and the majority carrier concentration is constant at the doping level. The minority carrier concentration is controlled by the transport equations 4.91 for electrons in the neutral p region and 4.92 for holes in the neutral n region. If recombination is linear then, in the neutral p region, $n(x)$ satisfies

$$\frac{d^2 n}{dx^2} - \frac{(n - n_0)}{L_n^2} + \frac{g(E, x)}{D_n} = 0 \qquad x < -w_p \qquad (6.15)$$

from Eq. 4.93 for $F = 0$, where $g(E, x)$ is the pair generation rate at x, given by

$$g(E, x) = (1 - R(E))\alpha(E) b_s(E) e^{-\alpha(E)x} \qquad (6.16)$$

from Eq. 4.25 for a uniform material. L_n is the electron diffusion length given by $L_n = \sqrt{\tau_n D_n}$ where τ_n is the electron lifetime. At the boundary with the depletion region $n(x)$ satisfies

$$n - n_0 = \frac{n_i^2}{N_a}(e^{qV/k_B T} - 1) \qquad x = -w_p \qquad (6.17)$$

where $n_0 = n_i^2/N_a$ is the *equilibrium* electron density in the p region. This boundary condition follows from the requirement that $E_{F_n} - E_{F_p} = qV_j$ throughout the space charge region (SCR). The second boundary condition is provided by the surface recombination at the outer surface

$$D_n \frac{dn}{dx} = S_n(n - n_0) \qquad \text{at } x = -x_p. \qquad (6.18)$$

We have used Eq. 4.85 and the fact that the electron current is purely diffusive when $F = 0$, hence

$$j_n(E, x) = qD_n \frac{dn}{dx}. \qquad (6.19)$$

For a p region which is thick compared with L_n the boundary condition $n(-x_p) = n_0$ is adequate.

In the neutral n region the hole population follows an analogous set of equations.

$$\frac{d^2p}{dx^2} - \frac{(p - p_0)}{L_p^2} + \frac{g(E, x)}{D_p} = 0 \qquad x > w_n \tag{6.20}$$

with the boundary conditions

$$p - p_0 = \frac{n_i^2}{N_d}(e^{qV/k_BT} - 1) \qquad \text{at } x = w_n \tag{6.21}$$

and

$$-D_P\frac{dp}{dx} = S_p(p - p_0) \qquad \text{at } x = w_n \tag{6.22}$$

where $p_0 = \frac{n_i^2}{N_d}$.

The hole current density obeys

$$j_p(E, x) = -qD_p\frac{dp}{dx}. \tag{6.23}$$

The majority carrier concentrations are simply

$$p = N_d \qquad x < -w_p \tag{6.24}$$

in the p region and

$$n = N_d \qquad x > w_n \tag{6.25}$$

in the n region.

6.4.2.   *Currents and carrier densities in the space charge region*

The space charge region is assumed free of carriers for the purpose of calculating the electrostatic potential. One approach is to ignore the SCR completely in the calculation of current. A better, though not self consistent, approximation is to calculate n and p within the space charge region from the Fermi levels and the electrostatic potential.

The depletion approximation requires that E_{F_n} and E_{F_p} are constant across the region, thus

$$\begin{aligned}
E_{F_n}(x) &= E_{F_n}(w_n) \\
E_{F_p}(x) &= E_{F_p}(-w_p) \\
qV &= E_{F_n}(x) - E_{F_p}(x) \qquad -w_p < x < w_n.
\end{aligned} \tag{6.26}$$

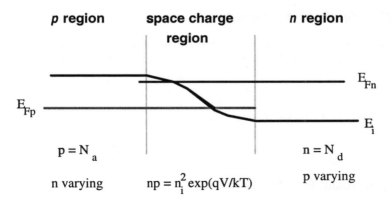

Fig. 6.4. Fermi levels and intrinsic potential in the depletion approximation. Within the space charge region n and p are determined by $E_{F_n} - E_i$ and $E_i - E_{F_p}$, respectively, and vary continuously from the majority carrier (doping) density at one side to the minority carrier density on the other.

From Eqs. 3.37 and 3.38 n and p are given by

$$n = n_i e^{(E_{F_n} - E_i)/k_B T}$$

and

$$p = n_i e^{(E_i - E_{F_p})/k_B T} .$$

E_i is found from $E_i = -q\phi$ where ϕ is the solution to Poisson's equation, given by Eqs. 6.8 and 6.9. Knowledge of E_{F_n}, E_{F_p} and E_i determines n and p completely within the SCR and ensures that n and p are continuous across the boundaries at $x = -w_n$ and $x = w_p$. Figure 6.4 shows how E_{F_n}, E_{F_p} and E_i vary across the region within the depletion approximation.

The depletion region current can be determined from carrier continuity. Integrating Eq. 4.87 across the SCR we find

$$J_{scr} = q \int_{-w_p}^{w_n} (U - G) dx \qquad (6.27)$$

where J_{scr} is equal to both the net electron current generated between $-w_p$ and w_n and to the net hole current generated between w_n and $-w_p$. J_{scr} can be broken down into spectral contributions, $j_{scr}(E)$, such that $J_{scr} = \int j_{scr}(E) dE$. U is known from n and p given the form of the recombination rate, and G is known from the incident spectrum and Eq. 6.16.

6.4.3.   *Total current density*

The net current is given by the sum of J_n and J_p at any point and is constant through the device in steady state. Evaluating J at $-w_p$ we find

$$J = -J_n(-w_p) - J_p(-w_p)$$

$$= -J_n(-w_p) - J_p(w_n) - J_{scr}$$

$$= -\int j_n(E, -w_p)dE - \int j_p(E, w_n)dE - \int j_{scr}(E)dE \qquad (6.28)$$

using the result, from Eq. 4.87, that

$$J_p(-w_p) = J_p(w_n) - \int_{-w_p}^{w_n} (U - G)dx \,.$$

$j_n(E, -w_p)$ and $j_p(E, w_n)$ are found by evaluating Eqs. 6.19 and 6.23 at the edges of the SCR using the solutions for $n(x)$ and $p(x)$. The result for J is identical if we evaluate the current at w_n.

We have chosen the sign convention here such that total current J is positive when flowing from p to n through the external circuit. This means that *photocurrent* is positive. This is opposite to the normal engineering convention, and is responsible for the minus signs in Eq. 6.28.

6.5.   General Solution for $J(V)$

Equation 6.15 for n has the general solution

$$n(E, x) = A_n \cosh\left(\frac{-x - w_p}{L_n}\right) + B_n \sinh\left(\frac{-x - w_p}{L_n}\right)$$

$$+ \gamma_n e^{-\alpha(x + x_p)} \qquad (6.29)$$

where

$$\gamma_n = \frac{\alpha(1 - R)b_s L_n^2}{D_n(\alpha^2 L_n^2 - 1)} \,. \qquad (6.30)$$

The boundary conditions, Eqs. 6.17 and 6.18, determine the coefficients A_n and B_n:

$$A_n = n_0(e^{qV/kT}) - \gamma_n e^{-\alpha(x_p - w_p)} \qquad (6.31)$$

and

$$B_n = \frac{\gamma_n[e^{-\alpha(x_p-w_p)}(\frac{S_n L_n}{D_n}\cosh\frac{(x_p-w_p)}{L_n} + \sinh\frac{(x_p-w_p)}{L_n}) - (\frac{S_n L_n}{D_n} + \alpha L_n)]}{- n_0(e^{qV/k_B T}-1)[\frac{S_n L_n}{D_n}\cosh\frac{(x_p-w_p)}{L_n} + \sinh\frac{(x_p-w_p)}{L_n}]}{\frac{S_n L_n}{D_n}\sinh\frac{(x_p-w_p)}{L_n} + \cosh\frac{(x_p-w_p)}{L_n}}.$$

(6.32)

From the definition of j_n, Eq. 6.19, we have

$$j_n(E, -w_p) = qD_n\left(-\frac{B_n}{L_n} - \alpha\gamma_n e^{-\alpha(x_p-w_p)}\right).$$

(6.33)

Hence,

$$J_n(-w_p) = \int j_n(E, -w_p)dE$$

with

$$j_n(E, -w_p)$$

$$= -\frac{q(1-R)b_s\alpha L_n}{(\alpha^2 L_n^2 - 1)}$$

$$\times \left\{ \frac{e^{-\alpha(x_p-w_p)}(\frac{S_n L_n}{D_n}\cosh\frac{(x_p-w_p)}{L_n} + \sinh\frac{(x_p-w_p)}{L_n}) - (\frac{S_n L_n}{D_n} + \alpha L_n)}{\frac{S_n L_n}{D_n}\sinh\frac{(x_p-w_p)}{L_n} + \cosh\frac{(x_p-w_p)}{L_n}} \right.$$

$$\left. + \alpha L_n e^{-\alpha(x_p-w_p)} \right\}$$

$$+ \frac{qD_n n_0(e^{qV/k_B T}-1)}{L_n}\left\{ \frac{\frac{S_n L_n}{D_n}\cosh\frac{(x_p-w_p)}{L_n} + \sinh\frac{(x_p-w_p)}{L_n}}{\frac{S_n L_n}{D_n}\sinh\frac{(x_p-w_p)}{L_n} + \cosh\frac{(x_p-w_p)}{L_n}} \right\}$$

(6.34)

For holes in the n region Eq. 6.20 for p has the general solution

$$p(E, x) = A_p\cosh\left(\frac{x+w_n}{L_p}\right) + B_p\sinh\left(\frac{x+w_n}{L_p}\right) - \gamma_p e^{-\alpha(x+x_p)}$$

(6.35)

where γ_p is given by Eq. 6.30 with L_n, D_n replaced by L_p, D_p and A_p and B_p are obtained from the boundary conditions 6.21 and 6.22 as

$$A_p = p_0(e^{qV/kT}-1) - \gamma_p e^{-\alpha(x_p+w_n)}$$

(6.36)

and

$$B_p = \frac{\gamma_p e^{-\alpha(x_p+w_n)}[(\frac{S_p L_p}{D_p}\cosh\frac{(x_n-w_n)}{L_p}+\sinh\frac{(x_n-w_n)}{L_p})-(\frac{S_p L_p}{D_p}-\alpha L_p)e^{-\alpha(x_n-w_n)}]}{\frac{S_p L_p}{D_p}\sinh\frac{(x_n-w_n)}{L_p}+\cosh\frac{(x_n-w_n)}{L_p}} - \frac{p_0(e^{qV/k_B T}-1)[\frac{S_p L_p}{D_p}\cosh\frac{(x_n-w_n)}{L_p}+\sinh\frac{(x_n-w_n)}{L_p}]}{\frac{S_p L_p}{D_p}\sinh\frac{(x_n-w_n)}{L_p}+\cosh\frac{(x_n-w_n)}{L_p}}.$$

(6.37)

This gives the solution for the hole current at w_n,

$$j_p(E, w_n) = -qD_p\left(\frac{B_p}{L_p} - \alpha\gamma_p e^{-\alpha(x_p+w_n)}\right)$$

(6.38)

hence

$$J_p(w_n) = \int j_p(E, w_n)\, dE$$

where

$$j_p(E, w_n)$$

$$= -\frac{q(1-R)b_s\alpha L_p}{(\alpha^2 L_p^2 - 1)}e^{-\alpha(x_p+w_n)}$$

$$\times\left\{\frac{(\frac{S_p L_p}{D_p}\cosh\frac{(x_n-w_n)}{L_p}+\sinh\frac{(x_n-w_n)}{L_p})-(\frac{S_p L_p}{D_p}-\alpha L_p)e^{-\alpha(x_n-w_n)}}{\frac{S_p L_p}{D_p}\sinh\frac{(x_n-w_n)}{L_p}+\cosh\frac{(x_n-w_n)}{L_p}} - \alpha L_p\right\}$$

$$+ \frac{qD_p p_0(e^{qV/k_B T}-1)}{L_p}\left\{\frac{\frac{S_p L_p}{D_p}\cosh\frac{(x_n-w_n)}{L_p}+\sinh\frac{(x_n-w_n)}{L_p}}{\frac{S_p L_p}{D_p}\sinh\frac{(x_n-w_n)}{L_p}+\cosh\frac{(x_n-w_n)}{L_p}}\right\}.$$

(6.39)

For each of Eqs. 6.34 and 6.39, the first term is proportional to b_s and is due to the light, and the second term is due to the electric bias. The light and bias induced currents are independent. This is a consequence of the linearity of the differential equations for n and p. Note also the difference in signs: the light generated and bias driven currents act in opposite directions.

The third contribution, J_{scr}, given by Eq. 6.27, is sometimes referred to as the recombination-generation current. It can be split into contributions from recombination, due to the applied bias

$$J_{rec} = q\int_{-w_p}^{w_n} U\, dx$$

(6.40)

and generation, due to the light,

$$J_{\text{gen}} = \int j_{\text{gen}}(E)dE$$

$$j_{\text{gen}}(E) = -q \int_{-w_p}^{w_n} g(E, x)dx \,. \tag{6.41}$$

The bias dependent and light dependent currents are, again, independent and opposite in sign. If the dominant recombination process is SRH recombination through trap states then, using Eq. 4.79 for U,

$$J_{\text{rec}} = -q \int_{-w_p}^{w_n} \frac{np - n_i^2}{\tau_n(p + p_t) + \tau_p(n + n_t)} dx \tag{6.42}$$

with n and p determined from E_{F_n}, E_{F_p} and E_i as explained above, and n_t and p_t given by Eqs. 4.72 and 4.77. The full form of the SRH recombination rate is used, rather than the linear approximation, because in this region both n and p vary strongly. In the depletion approximation E_{F_n} and E_{F_p} are constant across the SCR and E_i is a known function of x, given by Eqs. 6.8 and 6.9. If we make the approximation that E_i varies *linearly* across the depletion region then Eq. 6.42 can be evaluated analytically to yield

$$J_{\text{rec}}(V) = \frac{qn_i(w_n + w_p)}{\sqrt{\tau_n \tau_p}} \frac{2\sinh(qV/2kT)}{q(V_{\text{bi}} - V)/kT} \xi \tag{6.43}$$

where the factor ξ tends to $\pi/2$ at sufficiently large forward bias [Sah *et al.*, 1957].

The spectral contributions to the photocurrent, j_{gen} are found simply by integrating $g(E, x)$ across the SCR (Eq. 6.41). Thus

$$j_{\text{gen}}(E) = qb_s(1 - R)e^{-\alpha(x_p - w_p)}(1 - e^{-\alpha(w_p + w_n)}) \,, \tag{6.44}$$

expressing the fact the number of carriers generated is simply given by the number of photons absorbed. It is reasonable that carriers should be collected efficiently in the SCR on account of the built in field.

Taking Eqs. 6.43 and 6.44 into 6.27, and integrating Eq. 6.44 over photon energy we obtain,

$$J_{\text{scr}}(V) = \frac{qn_i(w_n + w_p)}{\sqrt{\tau_n \tau_p}} \frac{2\sinh(qV/2kT)}{q(V_{\text{bi}} - V)/kT} \frac{\pi}{2}$$

$$- q \int (1 - R)b_s e^{-\alpha(x_p - w_p)}(1 - e^{-\alpha(w_p + w_n)})dE \,. \tag{6.45}$$

Equations 6.28, 6.34, 6.39 and 6.45 give us the full solution for the current–voltage characteristic of the junction in the depletion approximation.

Below we consider how the solution simplifies in various special cases.

6.6. p–n Junction in the Dark

6.6.1.   *At equilibrium*

In equilibrium there is no illumination ($b_{\mathrm{s}} = 0$) and no applied bias ($V = 0$), so all contributions to the current density J are zero. The carrier populations obey

$$p(x) = N_{\mathrm{a}}, \quad n(x) = n_0 \qquad \text{for } x < -w_{\mathrm{p}} \tag{6.46}$$

$$p(x) = p_0, \quad p(x) = N_{\mathrm{d}} \qquad \text{for } x > -w_{\mathrm{n}} \tag{6.47}$$

and

$$pn = n_{\mathrm{i}}^2 \qquad \text{for } -w_{\mathrm{p}} < x < w_{\mathrm{n}} \,.$$

Clearly there will be no minority carrier diffusion currents in either neutral region and the fact that $pn = n_{\mathrm{i}}^2$ in the space charge region means that there is no net recombination. The bands and carrier profiles are illustrated in Fig. 6.5.

6.6.2.   *Under applied bias*

If a bias is applied to the junction in the dark, the built in potential barrier is reduced and more majority carriers are able to diffuse across the junction, so that there is now a net current of electrons from n to p and holes from p to n. This is sometimes referred to as the *injection* of majority carriers. Because the quasi Fermi levels in the SCR are split by the applied bias, there is net recombination in that region which adds to the current. The band profile is illustrated in Fig. 6.6.

In the dark $b_{\mathrm{s}}(E) = 0$ for all E and only the second term of each of Eqs. 6.34, 6.39 and 6.45 will be non zero. Because all of these contributions represent a current which flows in the direction of the bias, and in the opposite sense to the photocurrent it is convenient to define a dark current $J_{\mathrm{dark}}(V)$ which is positive,

$$J_{\mathrm{dark}} = J_{\mathrm{n}}(-w_{\mathrm{p}}) + J_{\mathrm{p}}(w_{\mathrm{n}}) + J_{\mathrm{scr}} \tag{6.48}$$

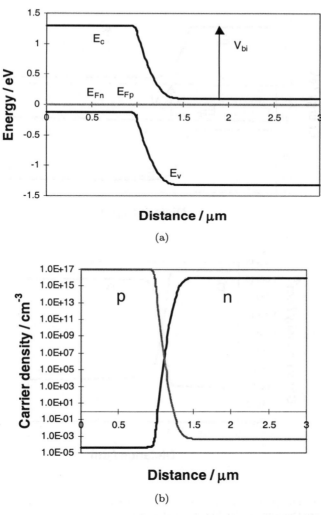

Fig. 6.5. (a) Band profiles and (b) carrier densities for a GaAs p–n junction in equilibrium. The parameters used were $x_p = 1$ μm, $x_n = 2$ μm, $N_a = 10^{17}$ cm^{-3}, $N_d = 10^{16}$ cm^{-3}. $L_n = L_p = 1$ μm and $S_n = S_p = 0$. Calculated using PC1D [Basore, 1991] as are Figs. 6.6, 6.8 and 6.11.

From Eqs. 6.34 and 6.39 it is clear that both the minority hole and minority electron currents are proportional to $(e^{qV/k_BT} - 1)$. In the limit where the neutral p layer is thick compared to L_n and the n layer thick compared to L_p, surface recombination becomes irrelevant and J_n and J_p have the

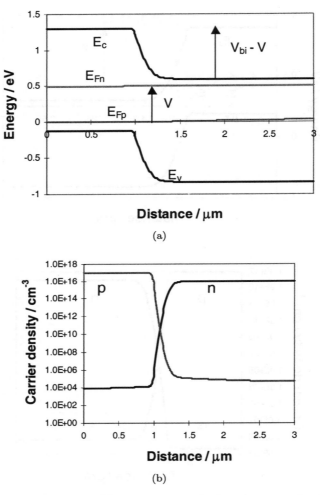

Fig. 6.6.   (a) Band profile and (b) carrier densities for biased p–n junction in the dark. Parameters are the same as for Fig. 6.5 and $V = 0.5$ V.

simple approximate forms

$$J_n(-w_p) = \frac{q n_i^2 D_n}{N_a L_n}(e^{qV/k_BT} - 1) \qquad (6.49)$$

and

$$J_p(w_n) = \frac{q n_i^2 D_p}{N_d L_p}(e^{qV/k_BT} - 1). \qquad (6.50)$$

In this limit, $p(x)$ and $n(x)$ have the simple forms

$$p - p_0 = p_0(e^{qV/K_BT} - 1)e^{-(x-w_n)/L_p} \tag{6.51}$$

for $x > w_n$ and

$$n - n_0 = n_0(e^{qV/k_BT} - 1)e^{(x+w_p)/L_n} \tag{6.52}$$

for $x < -w_p$ which express the fact that the mobile minority carrier density is significant only within a diffusion length of the depleted region. Further away, minority carriers are more likely to be lost by recombination before they reach the junction.

Since both J_n and J_p result from minority carrier diffusion they can be grouped together as a *diffusion* current, J_{diff}

$$J_{diff}(V) = J_n(-w_p) + J_p(w_n) = J_{diff,0}(e^{qV/k_BT} - 1) \tag{6.53}$$

where

$$J_{diff,0} = qn_i^2 \left(\frac{D_n}{N_aL_n} + \frac{D_p}{N_dL_p} \right). \tag{6.54}$$

Including the finite layer thickness and surface recombination velocity changes the *value* of J_{diff}, but not the bias dependence.

The recombination current from the depletion region given by 6.43 has the approximate form

$$J_{scr}(V) = J_{scr,0}(e^{qV/2k_BT} - 1) \tag{6.55}$$

where

$$J_{scr,0} = \frac{qn_i(w_n + w_p)}{\sqrt{\tau_n\tau_p}}. \tag{6.56}$$

Note that J_{scr} has a different dependence on both n_i and V to J_{diff}.

Combining terms we have

$$J_{dark}(V) = J_{diff,0}(e^{qV/k_BT} - 1) + J_{scr,0}(e^{qV/2k_BT} - 1)$$
$$+ J_{rad,0}(e^{qV/k_BT} - 1) \tag{6.57}$$

where we have included a term for the *radiative* recombination current, J_{rad}. From Eq. 4.57 it is clear that the radiative current has the same bias dependence as the diffusion current. J_{rad} may be significant in high quality, direct band gap materials. The relative importance of the terms depends on the importance of recombination in the SCR. In indirect gap materials like silicon, diffusion lengths are long compared to the depletion width and very little recombination occurs in the depleted region, so J_{scr} can be ignored. In that case we have

$$J_{dark} \approx J_{diff,0}(e^{qV/k_BT} - 1).$$ (6.58)

This is the Shockley or 'ideal' diode equation and is often quoted for the dark current of a solar cell.

In direct band gap materials where absorption is strong or where the SCR is wide, recombination within the depleted region may be dominant. Then

$$J_{dark} \approx J_{scr,0}(e^{qV/k_BT} - 1)$$ (6.59)

If more than one process is important then J_{dark} may appear to vary like

$$J_{dark} \approx J_0(e^{qV/mk_BT} - 1).$$ (6.60)

where J_0 is a constant and m is the *ideality factor*. An *ideal* diode has $m = 1$. m can be inferred from

$$\frac{1}{m} = \frac{kT}{q}\frac{d\ln J_{dark}}{dV}.$$ (6.61)

In real cells we often see a change in slope from J_{scr} dominating at low bias ($m = 2$), to J_{diff} or J_{rad} at high bias ($m = 1$) as shown in Fig. 6.7.

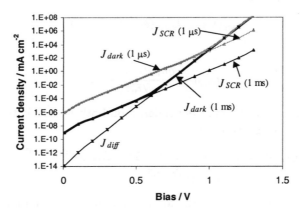

Fig. 6.7. Dark current of a non-ideal diode, resolved into contributions from diffusion and SCR recombination. As V increases the diffusion current overtakes the SCR recombination current and the slope of log J against V changes. The importance of the SCR current depends upon the carrier lifetimes: for shorter lifetimes, the transition to diffusive behaviour occurs at higher V than for longer lifetimes. Plots are shown for the cases where electron and hole lifetimes are both equal to 1 ms, and both equal to 1 μs.

In heterojunctions, tunnelling through barriers may give rise to m values greater than 2. The effect of an ideality factor greater than 1 is to reduce the fill factor of the solar cell. For ideal diodes the fill factor is 86%.

6.7.   *p–n* Junction under Illumination

6.7.1.   *Short circuit*

When the junction is illuminated, light creates electron-hole pairs in all three regions. n and p are then enhanced above their equilibrium values, and the electron and hole quasi Fermi levels are split. The electric field at the junction acts to separate the pairs by driving minority carriers across the junction. We will first obtain expressions for the photocurrent generated at *short circuit*, J_{sc}. In these conditions $V = 0$ so that, within the depletion approximation, E_{F_n} and E_{F_p} are equal across the SCR and there is no net recombination there. Figure 6.8 illustrates the band profile and carrier densities.

Now only the first term in each of Eqs. 6.34 and 6.39 is non-zero and all are negative, representing *positive* contributions to the photocurrent. For the minority carrier currents from the neutral regions we have

$$j_n(E, w_p)$$
$$= \left[\frac{qb_s(1-R)\alpha L_n}{(\alpha^2 L_n^2 - 1)}\right]$$
$$\times \left\{ \frac{(\frac{S_n L_n}{D_n} + \alpha L_n) - e^{-\alpha(x_p - w_p)}(\frac{S_n L_n}{D_n}\cosh\frac{(x_p - w_p)}{L_n} + \sinh\frac{(x_p - w_p)}{L_n})}{\frac{S_n L_n}{D_n}\sinh\frac{(x_p - w_p)}{L_n} + \cosh\frac{(x_p - w_p)}{L_n}} \right.$$
$$\left. - \alpha L_n e^{-\alpha(x_p - w_p)} \right\} \tag{6.62}$$

and

$$j_p(E, w_n)$$
$$= \left[\frac{qb_s(1-R)\alpha L_p}{(\alpha^2 L_p^2 - 1)}\right] e^{-\alpha(x_p + w_n)} \left\{ \alpha L_p \right.$$
$$\left. - \frac{\frac{S_p L_p}{D_p}(\cosh\frac{(x_n - w_n)}{L_p} - e^{-\alpha(x_n - w_n)}) + \sinh\frac{(x_n - w_n)}{L_p} + \alpha L_p e^{-\alpha(x_n - w_n)}}{\frac{S_p L_p}{D_p}\sinh\frac{(x_n - w_n)}{L_p} + \cosh\frac{(x_n - w_n)}{L_p}} \right\}. \tag{6.63}$$

The generation current in the SCR, j_{gen}, is given by Eq. 6.44.

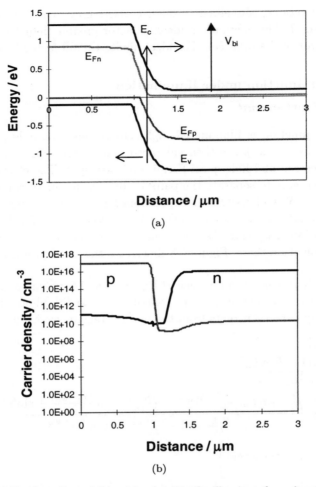

Fig. 6.8. (a) Band profile and (b) carrier densities for illuminated p–n junction at short circuit. Parameters are the same as for Fig. 6.5 and $b_s > 0$. The horizontal arrows in (a) show the direction of electron and hole transport following photogeneration.

Combining terms we have for the short circuit spectral photocurrent

$$j_{sc}(E) = -j_n(E, -w_p) - j_p(E, w_n) - j_{gen}(E). \qquad (6.64)$$

For a panchromatic spectrum the net photocurrent is

$$J_{sc} = \int_0^\infty j_{sc}(E)dE. \qquad (6.65)$$

The quantum efficiency of the cell at different photon energies E, defined in Chapter 1, is for some purposes of more interest than J_{sc} since it does not depend on the incident spectrum. We obtain $QE(E)$ easily from the spectral photocurrent since each component of $j(E)$ is proportional to $b_s(E)$:

$$QE(E) = \frac{-j_n(E, -w_p) - j_p(E, w_n) - j_{gen}(E)}{qb_s(E)}. \tag{6.66}$$

6.7.2. *Photocurrent and QE in special cases*

Let's consider the form of the photocurrent and the QE in some special cases.

In the case of a thick *p-n* cell where both the neutral layers are much thicker than the diffusion length the electron and hole currents can be approximated by

$$j_n(E, -w_p) \approx qb_s(1 - R)e^{-\alpha(x_p - w_p)} \left(\frac{\alpha L_n}{1 - \alpha L_n} \right) \tag{6.67}$$

and

$$j_p(E, w_n) \approx qb_s(1 - R)e^{-\alpha(x_p + w_n)} \left(\frac{\alpha L_p}{1 + \alpha L_p} \right). \tag{6.68}$$

Here only electrons generated within one diffusion length of the space charge region are collected. In the other extreme, where L_n and L_p are much longer than the p and n regions, respectively, we have

$$j_n(E, -w_p) \approx qb_s(1 - R) \left(1 + \frac{S_n}{\alpha D_n} \right) (1 - e^{-\alpha(x_p - w_p)}) \tag{6.69}$$

and

$$j_p(E, w_n) \approx qb_s(1 - R)e^{-\alpha(x_p + w_n)} \left(1 - \frac{S_p}{\alpha D_p} \right) (1 - e^{-\alpha(x_n - w_n)}). \tag{6.70}$$

In this limit, if no carriers are lost through recombination at the surfaces, ($S_n = 0$ or $S_p = 0$) the electron and hole photocurrents are simply q times the flux of photons absorbed in that layer.

The QE spectrum reflects the cell design and the material quality. For a *p-n* cell the short wavelength response is provided mainly by the p region, since high energy photons are absorbed at the front of the cell. Since carriers generated near the front surface are susceptible to surface recombination, the short wavelength QE is particularly sensitive to the surface recombination velocity. S_n may be reduced by introducing a window layer or passivating the surface. Long wavelength QE is affected by the back surface quality and by the thickness of the cell. The abruptness of the QE edge reflects the form of the absorption. At intermediate wavelengths carriers are generated in the SCR. The various contributions to the QE are shown in Fig. 6.9. The overall magnitude of the QE is affected by the efficiency of light absorption. It may be increased by reducing reflection losses, by increasing the width of the cell or by light trapping techniques.

Figure 6.10 shows the effect of varying some of the junction parameters. In Fig. 6.10(a) shows how the QE changes with emitter thickness. When x_p exceeds L_n, part of the emitter becomes a 'dead' layer which absorbs light but generates no photocurrent, and degrades the cell performance. It is straightforward to show that in the optimum design the emitter should be as thin as possible and certainly thinner than L_n. Practical solar cells are usually designed to have a thin, highly doped emitter and a thick, lightly doped base. In these conditions the photocurrent from the thin p region is likely to be limited by surface recombination velocity, while the current from the thick n region may be limited by either minority carrier diffusion length or by rear surface recombination. The effect of S_n on such a device is shown in Fig. 6.10(b) and the effects of L_p and S_p in Fig. 6.10(c).

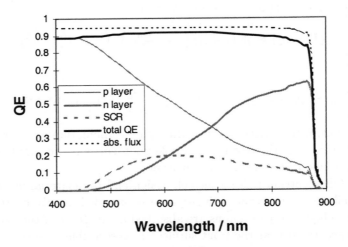

Fig. 6.9. Schematic quantum efficiency curve showing contributions from p, n, and SCR.

6.7.3. *p–n junction as a photovoltaic cell*

If a resistive load is connected between the terminals of the illuminated cell, the cell experiences both electrical and optical bias. The bias splits the Fermi levels throughout the device, as shown in Fig. 6.11. Notice how the

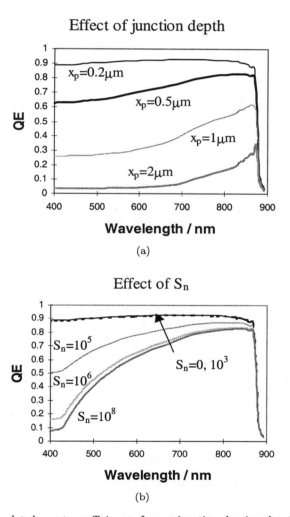

Fig. 6.10. Calculated quantum efficiency of a *p–n* junction showing the effect of (a) the emitter layer thickness (b) the front surface recombination velocity for a thin emitter cell, and (c) the transport parameters in the base for a thin emitter cell with low S_n. Surface recombination velocities are given in m s^{-1}.

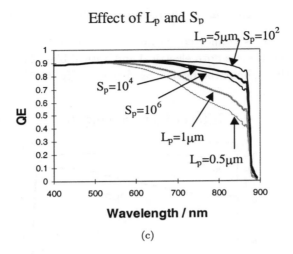

(c)

Fig. 6.10. (*Continued*).

Fermi levels slope towards the band edge at the edges of the SCR, as they did in the case of the short circuit cell, indicating that the net current is negative, *i.e.*, a photocurrent.

Now both terms in Eqs. 6.34 and 6.39 are non-zero. However, because the solutions for bias and light induced currents are independent in the depletion approximation, the solution for the current J in this regime is given by the algebraic sum of the integrated short-circuit photocurrent and the dark current at that bias (Eq. 1.4)

$$J(V) = J_{\mathrm{sc}} - J_{\mathrm{dark}}(V).$$

This superposition of currents is valid within the depletion approximation when minority carrier recombination is linear. If series resistance is important, the addition of terms changes slightly, as explained below.

If the dark current is dominated by diffusion currents (Eq. 6.58) then

$$J(V) = J_{\mathrm{sc}} - J_{\mathrm{diff},0}(e^{qV/k_{\mathrm{B}}T} - 1) \qquad (6.71)$$

with the dark saturation current $J_{\mathrm{diff},0}$ given by Eq. 6.54. This is identical to the ideal diode equation, Eq. 1.5, and is the most commonly used form for the current–voltage characteristic of a solar cell. More generally, the $J(V)$ can sometimes be expressed by the non-ideal diode form

$$J(V) = J_{\mathrm{sc}} - J_{\mathrm{m},0}(e^{qV/mk_{\mathrm{B}}T} - 1) \qquad (6.72)$$

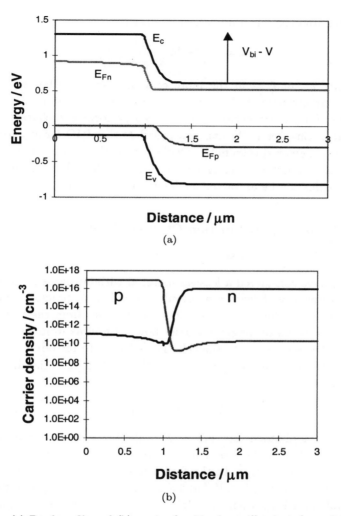

Fig. 6.11. (a) Band profile and (b) carrier densities in an illuminated *p–n* junction at positive bias. Notice that the carrier densities are similar to the values at short circuit (Fig. 6.8). This is because the bias chosen is much less than V_{oc} and the current is dominated by the photocurrent.

where $J_{m,0}$ is the saturation current in the dark. When recombination currents in the SCR are dominant, $m = 2$ and $J_{m,0} = J_{scr,0}$.

Equation 6.72 can be rearranged to find an expression for the open circuit voltage, V_{oc}, the voltage which develops between the terminals of the illuminated cell when no current is drawn

$$V_{oc} = \frac{mk_BT}{q} \ln\left(\frac{J_{sc}}{J_{m,0}} + 1\right). \qquad (6.73)$$

Equation 6.71, or 6.72 in the more general case, describes the current–voltage characteristic of the p–n junction solar cell. The relationship between the $J(V)$ characteristic and the cell performance characteristics is discussed in Chapter 1. We can now calculate the $J(V)$ for any p–n device structure in a given spectrum by calculating the QE spectrum and the dark current characteristic as described above. Then the solar cell performance characteristics can be extracted as described in Chapter 1.

6.8.   Effects on p–n Junction Characteristics

6.8.1.   *Effects of parasitic resistances*

So far we have considered a lossless material without parasitic resistances. Series resistance is, however, a prevalent problem in practical solar cells. As mentioned in Chapter 1, series resistance arises from the resistance of the p and n layers to majority carrier flow, as well as from the resistance of the electrical contacts to the cell. The effect is to introduce a potential drop of JAR_s between the junction and the applied potential difference at the terminals, V_{app},

$$V_{app} = V - JAR_s.$$

Thus, when a photocurrent flows ($J > 0$), the bias at the terminals is smaller than that at the junction, and when a dark current flows the bias at the terminals is larger. In the formulae derived above, V refers to the applied bias at the junction. If J and J_{dark} are both expressed in terms of the measurable bias V_{app}, then the additivity condition should be changed to

$$J(V_{app} + JAR_s) = J_{sc} - J_{dark}(V_{app} - J_{dark}AR_s)$$

to take account of series resistance losses. The effect of series resistance on $J(V)$ characteristics is illustrated in Fig. 1.10.

6.8.2.   *Effect of irradiation*

For linear photogeneration and recombination, increasing the intensity of the incident light increases the photocurrent. J_{sc} varies linearly with

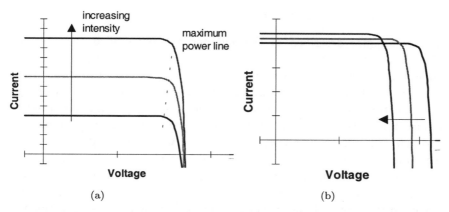

Fig. 6.12. Effects of (a) concentration and (b) temperature on p–n junction $J(V)$ characteristic. The arrow indicates the direction of increasing intensity or temperature.

concentration factor up to high irradiation levels. Equation 6.73 indicates that V_{oc} should also increase, logarithmically, with irradiation. Therefore, we expect light concentration to increase cell efficiency. The effect of light intensity on $J(V)$ is illustrated in Fig. 6.12.

In practice, concentration also increases series resistance and raises the temperature of the cell. These effects tend to degrade cell performance, with the result that the cell has an optimum efficiency at some finite concentration, usually a few hundred suns. This will be discussed further in Chapter 9.

6.8.3. *Effect of temperature*

As temperature is increased, the equilibrium population of electrons n_i increases exponentially, increasing the dark saturation current density. Note that this effect will be stronger for the diffusion component of the dark current than for the recombination-generation component, because of the stronger dependence on n_i (Eqs. 6.54 and 6.56). The increased dark current reduces V_{oc}. At the same time, increased temperature reduces the band gap and increases the photocurrent, since lower energy photons can now be absorbed. The net effect is a reduction in efficiency because the loss in V_{oc} outweighs the gain in J_{sc}. This is illustrated in Fig. 6.12(b).

6.8.4. *Other device structures*

The techniques described in this chapter can be applied to other device structures, provided that the depletion approximation is valid. More complicated structures may include 'window' layers placed in front of the emitter; highly doped back surface field layers; graded p or n layers; p–i–n structures and heterojunctions. A wide band gap window is used to reduce the surface recombination velocity and to modify light admission through reflection and absorption of short wavelength light. These effects are easily incorporated in the model as changes to S_n and $b_s(E)$. Modelling a back surface field layer, such as n^+ back surface layer beneath an n type base, means solving for the n–n^+ junction and finding minority carrier diffusion currents in both regions. A compositionally graded p or n layer can be used to assist the minority carrier flow towards the junction and can be treated by including the electric field term in the current equations. A p–i–n structure is straightforward, and can be treated by extending the depletion region to include an undoped layer. This is discussed in Chapter 8 for amorphous silicon. A p–n heterojunction may be treated just as the p–n junction, once the heterojunction potential profile is worked out, and the difference in absorption coefficient included.

6.8.5. *Validity of the approximations*

The depletion approximation is best for highly doped junctions at low bias. Then the rectangular form assumed for the space charge profile is most accurate, and the volume taken up by the junction is the smallest. For weakly doped junctions or at high forward bias, free carriers populate the junction region and contribute to the potential distribution. The effect is generally to increase recombination, since the electron and hole densities are more likely to be similar, and the Shockley Read Hall recombination rate larger. Thus the depletion approximation underestimates the dark current at low doping or high forward bias.

Linear recombination results for either Shockley Read Hall or radiative recombination in the limit where the majority carrier density greatly outnumbers the minority carrier. This will be untrue for lightly doped materials or in the limit of high light intensity. Some other recombination processes are not linear, even in the limit of low minority carrier density. In some defective materials the recombination rate may rise superlinearly with minority carrier density because of the effects of traps. In such conditions, light and dark currents may not be added, and the light current may even vary nonlinearly with light intensity.

6.9. Summary

At a p–n junction between a p type and n type layer of the same semiconductor, a potential barrier is formed which acts as a barrier to majority carrier flow. This barrier provides the asymmetric resistances needed for photovoltaic action. Good quality p–n junctions can be made easily and they are the most widely used design of solar cell.

The current–voltage characteristic for a p–n junction solar cell can be calculated analytically by making two approximations. These are that the potential is dropped entirely across a junction region which is free of carriers (the depletion approximation), and that the minority carrier recombination rates are linear in the carrier density. Then the junction may be treated in three layers, neutral n and p type layers and a charged space charge region. The current is due to contributions from each of the layers, which are calculated separately. In each of the neutral layers the current is due to minority carriers and can be found by solving the current and continuity equations. These equations are simplified when the electric field is zero and recombination linear. In the space charge region the current is due to generation and recombination between both types of carrier. The net current voltage characteristic is the sum of contributions due to the light (the photocurrent) and due to the applied bias (the dark current). The dark currents from the neutral layers and space charge layers have different bias dependence. $J(V)$ can be expressed in the form of the Shockley diode equation with an ideality factor which expresses the dominant type of dark current. The photocurrent can be related to the quantum efficiency spectrum, and is influenced by the junction design and material parameters. Photocurrent is highest when the greatest amount of light results in carriers crossing the junction. This is true for a thin emitter, low front surface recombination velocity, long diffusion length in the base and low reflectivity. Dark current is least when carrier lifetimes are the longest.

The formulation can be used to simulate the effects of junction design, materials properties, irradiance and other parameters on solar cell performance, and is easily modified to model other varieties of p–n junction.

References

C.-T. Sah, R.N. Noyce and W. Shockley, "Carrier generation and recombination in $p–n$ junctions and $p–n$ junction characteristics", *Proc. Institute of Radio Engineers*, 1228 (1957).

P.A. Basore, "PC1D version 3: improved speed and convergence", *Conf. Record. 22nd IEEE Photovoltaic Specialists Conf.*, 299–302 (1991). PC1D software was developed by Sandia National Laboratories, NM, USA; and is distributed by Centre for Photovoltaic Engineering, University of New South Wales.

Chapter 7

Monocrystalline Solar Cells

7.1. Introduction: Principles of Cell Design

The most common solar cell design is a monocrystalline p–n junction. p–n homojunctions have the significant advantage over heterojunction designs that there is no material interface at the junction, and so losses due to interface states can be avoided. In this chapter we will discuss the design of p–n junction solar cells in the two best performing photovoltaic materials, monocrystalline silicon and gallium arsenide, illustrating the issues relevant to PV in weakly absorbing and strongly absorbing materials. We will look at the techniques which are used to improve performance in practice.

Efficient photovoltaic energy conversion requires efficient light absorption, efficient charge separation and efficient charge transport. From Chapter 6 we can translate these requirements into demands on the material and design parameters of a p–n junction solar cell.

- For good optical absorption, the optical depth of the device, $\alpha(E)(x_p + x_n)$ should be high for energies E above the band gap, and the reflectivity of the surface, $R(E)$, should be small.
- For good charge separation, the built in bias V_{bi} should be large, demanding a high doping gradient across the junction (a large $N_a \times N_d$ product). Charge recombination in the junction region should be as slow as possible. And, as we have seen in Chapter 6, the junction should be located close to the surface for effective charge separation over a range of wavelengths.
- For efficient minority carrier transport the minority carrier lifetimes (τ_n, τ_p) and diffusion lengths (L_n, L_p) should be long and surface recombination velocities (S_n, S_p) should be small. Good majority carrier transport requires small series resistance R_s, and the shunt resistance R_{sh}

should be as high as possible to avoid leakage of carriers back across the junction.

- The band gap should be close to the optimum for the intended solar spectrum, $b_s(E)$.

Some of these demands are contradictory and a compromise must be found. For instance, high doping levels needed for a large built in bias are likely to reduce minority carrier lifetimes, and strong optical absorption will generally mean strong recombination.

7.2.   Material and Design Issues

7.2.1.   *Material dependent factors*

There are certain basic materials requirements. For high theoretical conversion efficiencies in AM1.5 ($\geq 30\%$) the band gap should lie in the range 1–1.6 eV; or in the range 0.7 to 2 eV for efficiencies over 20% (see Fig. 7.1). p and n type varieties of the semiconductor material must be available, and we need to be able to produce a junction of good quality. For acceptable photocurrent levels the quantum efficiency should be high over a broad range of wavelengths. That means that the diffusion lengths must be long compared to the absorption depth, and this places high demands on the crystal quality.

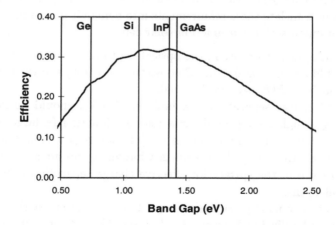

Fig. 7.1.   Limiting efficiency as a function of band gap for the air mass 1.5 solar spectrum, and the band gaps of some common photovoltaic materials.

These requirements limit the range of materials which can be used. Figure 7.1 shows the band gaps of some common semiconductor materials compared to the limiting efficiency in a standard solar spectrum. Of the elemental semiconductors, only silicon has a suitable band gap. Germanium and selenium were used in early photovoltaic cells but the band gaps are too small. Of the compound semiconductors, III–V crystalline materials have been best developed. Gallium arsenide and indium phosphide have suitable band gaps. Several II–VI and I–III–V semiconductors are strong optical absorbers with suitable band gaps and some, notably CdTe and CuInSe$_2$, have been developed for photovoltaics. These materials cannot always be doped both n and p type, so that p–n homojunctions cannot be made. Amorphous silicon and related alloys have suitable band gaps, but diffusion lengths are too short compared to the absorption depth, and so again they cannot be used for p–n homojunction cells.

Materials can be prepared as single crystals (monocrystalline) or as polycrystalline or multicrystalline wafers. If grain sizes are smaller than device thickness, grain boundary effects are likely to dominate transport and diffusion lengths are likely to be short. In such cases, much of the analysis of Chapter 6 is not valid. We will distinguish 'multicrystalline' as material where the crystal grain size is comparable with or larger than the device thickness, and 'polycrystalline' or 'microcrystalline' as material where the grain size is much smaller. Devices made from multicrystalline material can be considered in same way as single crystal devices, treated below. Junctions in polycrystalline materials will be considered in the Chapter 8.

Intrinsic materials variables are the optical properties, the available dopants, the carrier mobilities, and the dominant recombination processes, all of which are related to the crystal structure. Process variables are the means of wafer and junction fabrication.

7.2.2. *Design factors*

The requirement of a good quality junction generally favours the p–n homojunction for the design. Once the material is chosen, a limited number of design factors remain which can be controlled. These include junction polarity; junction depth; doping levels; doping gradients; cell thickness; surface treatments; and contact design. The simple p–n junction can also be modified in a number of ways to improve performance. Strategies include compositional variations such as windows and graded layers, and light trapping structures.

The main focus of cell design depends upon the optical properties of the material. In a weakly absorbing material like silicon, strategies to maximise the optical absorption are important, and recombination at the rear surface needs to be avoided. In a highly absorbing material like gallium arsenide, good optical absorption is achieved easily but more attention must be paid to minimising recombination near the front surface and around the junction. Strategies for these two groups of materials are discussed below.

7.2.3. General design features of p–n junction cells

A number of general design features apply to p–n homojunction cells independent of the material. These are that:

 (i) the thickness should exceed the absorption length, for efficient light absorption;
 (ii) the junction should be shallow compared to both the diffusion length in the emitter and the absorption length, to avoid having a 'dead layer' at the front of the cell where light is absorbed unproductively, as noted in Chapter 6;
 (iii) the emitter should be doped heavily, to improve conductivity to the fine metallic contacts on the front of the cell. Heavy emitter doping also allows the base to be doped lightly, which improves collection in the neutral base region without limiting the open-circuit voltage.
 (iv) Reflection of light should be minimised. All crystalline semiconductor cells are treated with an anti-reflection (AR) coat. This is a thin layer of material of refractive index between that of the semiconductor and the air which couples light of preferred wavelengths into the semiconductor by matching the optical impedance. The AR coat material and thickness are chosen to maximise photocurrent generation for the relevant incident spectrum. (This is discussed further in Chapter 9.)

7.3. Silicon Material Properties

7.3.1. Band structure and optical absorption

Silicon is a group IV element which adopts the tetrahedral crystal structure at room temperature and pressure. In this arrangement every valence electron is involved in bonding and the material is a semiconductor. As discussed in Chapter 3, the band gap is indirect. That is, although the lowest point in the conduction band and the highest point in the valence band

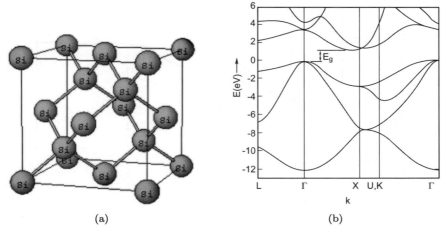

(a)   (b)

Fig. 7.2. (a) Tetrahedral crystal structure of crystalline silicon; (b) Band structure of silicon, showing energy levels as function of wavevector, **k**. Levels are normally filled up to the energy $E = 0$, which marks the top of the valence band. Notice that the valence band maximum is at the Γ point, where $\mathbf{k} = 0$, while the conduction band minimum is near the X point, where **k** is directed along (100). The fundamental band gap lies between these extrema and is marked as E_g.

— the fundamental band gap — are separated by only 1.1 eV at room temperature, these points occur at different values of the crystal momentum and a phonon is needed as well as a photon in order for an electron to be promoted. This reduces the optical absorption compared to direct gap semiconductors. *Direct* optical transitions occur in silicon at photon energies above 3 eV though these are not useful for photovoltaics.

The band gap is fairly close to the optimum for solar energy conversion. At 1.1 eV, the theoretical limiting efficiency in AM1.5 is a few percent smaller than the maximum.

Silicon has a refractive index of around 3.4 and a natural reflectivity of about 40% over visible wavelengths. Introducing a single or multilayer anti-reflection coat reduces this to less than 5%.

7.3.2. Doping

Another criterion for electronic materials is whether they can be easily doped. Silicon can be readily doped n type, usually by the addition of pentavalent phosphorus impurity atoms, which introduce an extra loosely bound electron when substituted for a tetravalent silicon atom. p type

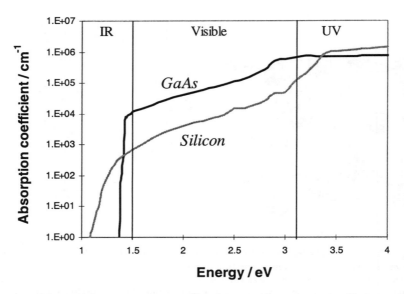

Fig. 7.3. Silicon absorption spectrum. The absorption spectrum of gallium arsenide is shown for comparison, to emphasise the low absorption in silicon due to its indirect band gap. The vertical lines denote the wavelengths 800 nm and 300 nm, the approximate limits of the visible part of the spectrum.

silicon may be produced by doping with trivalent boron, which introduced an electron vacancy, or hole, into the lattice when substituted for silicon. Addition of these impurity atoms generally degrades the material quality. However relatively high doping levels are needed in the emitter to reduce series resistance and increase V_{bi}, which in turn increases V_{oc}. In silicon the influence of doping on V_{oc} is limited by the shrinkage of the band gap, due to the introduction of tail states, at high doping levels. This means that V_{oc} cannot in fact exceed 81% of E_g at room temperature.

7.3.3.   Recombination

We will see below that nearly all of the volume of a typical crystalline silicon cell is provided by p type silicon. The effect of recombination in the depletion region and emitter layer is rather low because of the low levels of photo-generation in those thin layers, and can usually be neglected. This leaves electron recombination in the p region as the dominant volume recombination process. This includes radiative, Auger and trap-assisted mechanisms.

The net recombination rate, U, will be the sum of all processes,

$$U = U_{\text{rad}} + U_{\text{Aug}} + U_{\text{SRH}} . \tag{7.1}$$

From Eq. 4.59 we expect the radiative recombination rate in p type material to follow

$$U_{\text{rad}} = \frac{n - n_0}{\tau_{\text{n,rad}}} \tag{7.2}$$

where the radiative lifetime $\tau_{\text{n,rad}}$ varies like $1/N_a$. Since silicon is an indirect band gap semiconductor, radiative recombination is slow and $\tau_{\text{n,rad}}$ is typically of the order of milliseconds, so this contribution is negligible.

Auger recombination can be important in silicon, if the carrier density is high. For electrons in p type material, from Eq. 4.67

$$U_{\text{Auger}} = A_n N_a^2 (n - n_0) \tag{7.3}$$

so that the lifetime varies like

$$\tau_{\text{n,Aug}} \infty N_a^{-2} . \tag{7.4}$$

Auger processes are most important in silicon at high doping densities (over 10^{17} cm^{-3}) and can be the dominant recombination pathway mechanism.

Trap assisted nonradiative processes dominate in all but the purest silicon. For Shockley Read Hall recombination through a single trap state we expect, from Eq. 4.80,

$$U_{\text{SRH}} = \frac{n - n_0}{\tau_{\text{n,SRH}}} \tag{7.5}$$

where the minority carrier lifetime $\tau_{\text{n,SRH}}$ depends upon the density of trap states, as well as their position within the band gap. From Eq. 4.69 we know that

$$\tau_{\text{n,SRH}} = \frac{1}{B_n N_t} \tag{7.6}$$

and there is evidence that $\tau_{\text{n,SRH}}$ decreases as doping increases, suggesting that the number of active traps increases with doping. Lifetimes vary from ms for very pure material to a fraction of a μs for high defect densities.

Combining the bulk recombination processes in Eq. 7.1 and noting that in the neutral p region all recombination mechanisms are linear in the excess electron density, $(n - n_0)$, we obtain the net electron lifetime τ_n such that

$$U = \frac{n - n_0}{\tau_n} \tag{7.7}$$

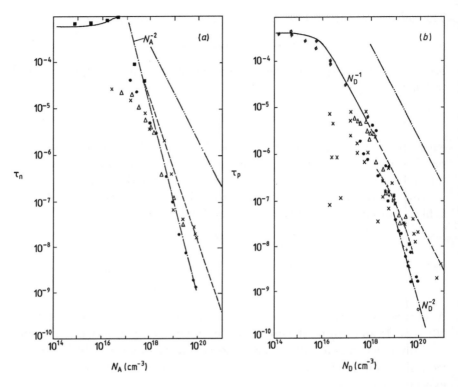

Fig. 7.4. Lifetime as a function of doping density in p type (left panel) and n type (right panel) silicon. At high doping densities the lifetime varies like the inverse square of the doping density, indicating that Auger processes are dominant. Lifetimes for other recombination processes are expected to vary like the simple inverse of the doping. From van Overstraeten and Mertens [van Overstraeten, 1986].

and

$$\frac{1}{\tau_n} = \frac{1}{\tau_{\text{SRH}}} + \frac{1}{\tau_{\text{rad}}} + \frac{1}{\tau_{\text{Aug}}} . \qquad (7.8)$$

Different processes will have different temperature and doping dependence. In lightly doped p-type silicon at room temperature, SRH processes dominate, giving rise to a τ value of about 10 μs. Lifetime in lightly doped n type material is rather shorter, around 1 μs. In more heavily doped silicon or at higher temperatures Auger dominates. This is evident from the dependence of τ on N_a shown in Fig. 7.4. Radiative lifetimes in silicon are extremely long and never dominate the recombination in a practical solar cell.

As we saw in Chapter 6, surface recombination is important when bulk recombination is low, *i.e.*, when minority carrier diffusion lengths are long. For electrons in p type silicon, the surface recombination velocity at untreated surfaces, and at interfaces with metallic contacts, is in the range of 10^3–10^5 cm s^{-1} [Hovel, 1975]. When the surface is passivated with a layer of silicon dioxide, the oxide shields minority carriers from defects at the surface and reduces S_n to less than 100 cm s^{-1} [Green, 1995].

7.3.4. *Carrier transport*

In practice, the electron mobility in p type silicon tends to be higher than the hole mobility in n type silicon doped to the same level, and the minority electron diffusion length is longer. This is because the fundamental impurities in silicon are of acceptor type and so are more important in n type than p type material. It means that carrier collection is more efficient in p type material than in an n type layer of the same doping density and thickness, and so cells are generally designed as n–p cells with a thin n type emitter on top of a thick p type base.

Electron and hole mobilities are determined by the frequency of scattering events within the conduction or valence band. At low doping levels scattering is dominated by the silicon lattice. At high doping mobility decreases because the impurity atoms generate scattering centres. To a first approximation the mobility does not depend on whether doping is n type or p type. In silicon, electrons have a mobility of around 1500 cm^2 V^{-1} s^{-1} at low doping, falling to around 70 cm^2 V^{-1} s^{-1} at impurity concentrations of over 10^{19} cm^{-3}, while holes have a low-doping mobility of 500 cm^2 V^{-1} s^{-1}, falling to around 50 cm^2 V^{-1} s^{-1}. The corresponding diffusion constants in lightly doped material are 30–40 cm^2 s^{-1} for electrons and around 10 cm^2 s^{-1} for holes [Shur, 1990; Green, 1995].

Minority carrier diffusion lengths are defined from $L = \sqrt{\tau D}$ (Eq. 4.95) where τ refers to the *net* recombination and do not distinguish between processes. For a commercial silicon cell, τ_n is several μs and L_n is around 100 μm.

The resistivity of any semiconductor material, ρ, depends upon the level of doping. From the definition of conductivity (Eq. 3.41),

$$\rho = \frac{1}{\sigma} = \frac{1}{q(\mu_n n + \mu_p p)} \, . \tag{7.9}$$

In doped material, resistivity will therefore be dominated by the majority carrier density and mobility. ρ varies roughly like $(10^{16}/N_a)$ Ohm-cm in p type silicon, and is a factor of three smaller in n type. With an acceptor concentration of 10^{16} cm^{-3}, a p region several hundred μm thick has a sheet resistance of around 10^{-1} Ohm-cm^2 which is unimportant for the current densities under standard solar illumination. However, series resistance effects can be more serious in the emitter, where the funnelling of current from a large semiconductor area into a very small contact area creates high current densities. This is one reason why the emitter is highly doped relative to the base. When the contact resistance is included the equivalent series sheet resistance for a commercial silicon cell is typically 20–50 Ohm-cm^2.

Silicon materials properties are discussed in detail by [Green, 1995; Bube 1998; van Overstraeten, 1986]; and many semiconductor texbooks.

7.4. Silicon Solar Cell Design

7.4.1. *Basic silicon solar cell*

A typical silicon solar cells is an n–p junction made in a wafer of p type silicon a few hundred microns thick and around 100 cm^2 in area. The p type wafer forms the *base* of the cell and is thick (300–500 μm) in order to absorb as much light as possible, and lightly doped ($\sim 10^{16}$ cm^{-3}) to improve diffusion lengths. The n type emitter is created by dopant diffusion and is heavily doped ($\sim 10^{19}$ cm^{-3}) to reduce sheet series resistance. This layer should be thin to allow as much light as possible to pass through to the base, but thick enough to keep series resistance reasonably low. Carrier collection from the emitter is negligible because of high recombination in this heavily doped layer. The front surface is anti-reflection coated and both front and back surfaces are contacted before encapsulation in a glass covering.

7.4.2. *Cell fabrication*

Single crystal silicon may be grown by a number of methods. In the common *Czochralski* process a single crystal is drawn slowly out of a melt. In the *float zone* process a single crystal is gradually formed from a polycrystalline rod by passing a molten zone through it. This is more costly but produces higher purity material. In either case the dopant (usually boron) is introduced during growth to produce a p type crystal. The solid crystal is sliced into wafers and etched to smooth the rough surfaces.

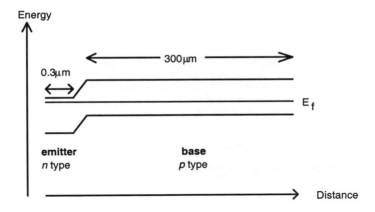

Fig. 7.5.   Energy band diagram of a *n–p* junction in silicon.

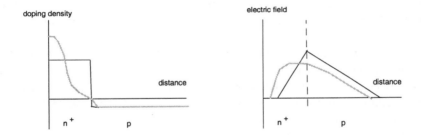

Fig. 7.6.   Doping profiles for an abrupt (black line) and diffused (grey line) junction.

The junction is prepared by diffusing the *n* type dopant — usually phosphorus — on to the *p* type wafer. Phosphorus may be deposited either from the vapour phase by exposure to nitrogen gas bearing POCl₃; from solid phase, for example by chemical vapour deposition of phosphorus oxide; or directly by ion implantation. The latter method allows greater control of the doping profile but is more costly. Note that the dopant profile in the diffused *n* layer will not be uniform and the junction not abrupt, unlike the ideal *p–n* junction in Chapter 6. This has the consequence that the electric fields extend further from the junction (and the depletion approximation is less accurate). The diffused doping profile may be modelled with an error function.

Multicrystalline silicon, which is used in most commercial silicon cells, is made by a variety of methods such as casting and ribbon growth. (For

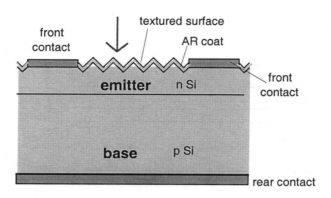

Fig. 7.7.   Layer structure of basic silicon cell.

details see Green [Green, 1995].) The relatively large sizes of the grains (0.1–10 cm) mean that moderately efficient devices can be prepared from multicrystalline material using techniques similar to those used for monocrystalline silicon.

The front surface is usually textured to reduce reflectivity and an antireflection coating is deposited from liquid or vapour phase added. For silicon the AR coating should have a refractive index of around 2 and thickness of 80–100 nm. Suitable materials for silicon are tantalum oxide (Ta_2O_5), titania (TiO_2) and silicon nitride (Si_3Ni_4).

The rear surface is doped more heavily to create a back surface field, which helps to reduce the loss of carriers through surface recombination. This is discussed below.

Finally the front and back contacts are added. In the early silicon cells, aluminium was used as the rear contact. In large scale production, AR coat, front and back contacts are usually deposited by screen printing and then fired. Screen printing of contacts is cheap but obscures a relatively large area of the cell and degrades conductivity.

7.4.3. *Optimisation of silicon solar cell design*

Figure 7.8 illustrates the absorption and recombination profiles in a typical silicon cell, calculated using the p–n junction theory of Chapter 6 and the materials parameters in Table 7.1. From the graphs we can make the following observations:

• absorption of light close to the band gap (near infrared) is poor

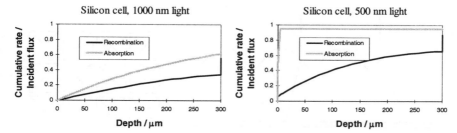

Fig. 7.8. Simulated cumulative absorption and recombination rates for a silicon solar cell with the parameters in Table 7.1 under monochromatic light at two different wavelengths at a bias close to open circuit. In each graph the cumulative absorption and recombination, normalised to the total incident flux, are plotted against depth through the cell. For a good cell at short circuit, recombination should be zero and cumulative absorption should be close to one. At open circuit (illustrated) the final values of recombination and absorption are equal.

Table 7.1. Typical silicon material and cell parameters, used for the calculations in Fig. 7.8.

	EMITTER (N TYPE)	BASE (P TYPE)
thickness/μm	0.5	300
doping/cm^{-3}	1×10^{19}	1×10^{16}
minority carrier diffusion constant/cm^2 s^{-1}	2	40
minority carrier lifetime/s	1×10^{-6}	5×10^{-6}
minority carrier diffusion length/μm	14	140
surface recombination velocity/cm s^{-1}	10000	10000
reflectivity	0.05	
absorption: at 500 nm/cm^{-1}	15000	
at 1 μm/cm^{-1}	35	

- bulk recombination in the p region is the most important recombination process
- rear surface recombination is important, particularly for photogeneration by red and infrared light
- front surface recombination and recombination in the junction region are relatively unimportant for photogeneration by long wavelengths

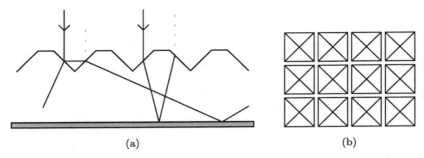

Fig. 7.9. Surface texturing: (a) shows how the refraction of incident light rays at the textured surface increases the optical path length relative to light normally incident on a plane surface; (b) shows a plan view of regular pyramids.

To improve the performance of the cell it is necessary to maximise the absorption of red light, minimise recombination at the rear surface, and minimise series resistance. Bulk recombination is determined mainly by the method of wafer growth, and for good quality silicon it is already as low as can be expected.

Thus the main challenges in crystalline silicon cell design are to:

- Maximise absorption
- Minimise rear surface recombination
- Minimise series resistance

These are discussed below.

7.4.4. Strategies to enhance absorption

- Texturing of front surface. This reduces the net reflection of light and increases the optical depth of the cell. Texturing can be achieved by treating with an anisotropic chemical etc. which acts preferentially along the (111) crystal planes and leaves a pattern of pyramids on the surface. Regular pyramids can be produced on a monocrystalline surface by photolithographic definition. Light trapping is improved by using *inverted* pyramids, which improve the total internal reflection of light reflected from the back surface, by asymmetric pyramids, or by texturing the rear surface. Light trapping strategies are discussed further in Chapter 9.
- Optimisation of contacts. Shading of the front surface by metal contacts reduces the surface area available to the incident light by as much as 10%. Reduced contact area increases the available surface area but increases

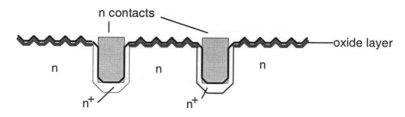

Fig. 7.10. Buried contacts. Differential n^+ doping in the contact grooves reduces series resistance.

the resistance either in the emitter, if the contacts are too sparse, or in the metal, if the fingers are too narrow. The optimum arrangement is a grid of narrow, dense, highly conducting fingers. One solution is to use narrow, deep contacts partly buried in the surface of the cell. This may reduce shading to a fraction of a percent of the surface. By embedding the contacts in the semiconductor, a larger contact area can be achieved without increasing the surface shading. The grooves are created by laser or mechanical etching, and are doped more heavily than the main emitter to improve conductivity. However, the large scale preparation of such contacts is more costly than screen printing.

7.4.5. Strategies to reduce surface recombination

- Back surface field. A more heavily doped layer is formed at the back surface of the p type base by alloying with aluminium or by diffusion. This introduces a p^+–p junction and presents a potential barrier to the minority electrons. This *back surface field* reflects electrons and reduces the effective rear surface recombination velocity, to less than 100 cm s^{-1}. (See Box 7.1). The extra p^+–p junction also adds to the built in bias of the cell, and may enhance V_{oc} [Hovel, 1975]. Front surface fields have also been used in some cell designs, but are less effective since the ratio of doping levels will be smaller.

Box 7.1. Calculation of effective surface recombination velocity with a back surface field

When a back surface field layer is present in p type material, the change in doping introduces a step in the conduction band edge of height $\Delta E_c =$

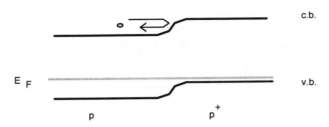

Fig. 7.11. Back surface field in p type material. Electrons are reflected from the doping barrier, effectively reducing the surface recombination velocity.

$k_B T \ln(\frac{N_{a+}}{N_a})$. Electron concentration is reduced on the p^+ side of the low-high junction (x_{j+}) relative to the main p side (x_{j-}) by the ratio of the doping levels, N_a/N_{a+}. This follows from continuity of the Fermi level across the junction. If $x = x_j$ at the p–p^+ junction we have

$$\Delta n(x_{j-}) = \Delta n(x_{j+}) \times e^{\Delta E_c/k_B T} = \Delta n(x_{j+}) \frac{N_{a+}}{N_a}$$

for the excess electron concentration Δn. Current continuity at the junction provides a second boundary condition,

$$D_{n-} \frac{dn(x_{j-})}{dx} = D_{n+} \frac{dn(x_{j+})}{dx}$$

where D_{n-} and D_{n+} are the diffusion constants in the p and p^+ layers. Now, if we define an effective surface recombination velocity at the junction, S_{eff}, in analogy with Eq. 4.85,

$$S_{eff} = -\frac{D_{n-}}{\Delta n(x_{j-})} \frac{dn(x_{j-})}{dx} ,$$

it follows that

$$S_{eff} = -\frac{D_{n+}}{\Delta n(x_{j+})} \frac{N_a}{N_{a+}} \frac{dn(x_{j+})}{dx} .$$

The electron density in the highly doped layer is fixed by the boundary condition at the rear surface

$$-D_{n+} \frac{dn(x_j + x_b)}{dx} = S_{n+}(\Delta n(x_j + x_b)) ,$$

where x_b is the thickness of the highly doped layer and S_{n+} is the rear surface recombination velocity within the highly doped layer. If recombination is linear in that layer and characterised by a diffusion length L_{n+} we can solve for the ratio $\frac{\frac{dn}{dx}(x_{j+})}{\Delta n(x_{j+})}$ and find, in the dark, that

$$S_{\text{eff}} = \frac{D_{n+}}{L_{n+}} \frac{N_a}{N_{a+}} \left\{ \frac{\frac{S_{n+}L_{n+}}{D_{n+}} \cosh \frac{x_b}{L_{n+}} + \sinh \frac{x_b}{L_{n+}}}{\frac{S_{n+}L_{n+}}{D_{n+}} \sinh \frac{x_b}{L_{n+}} + \cosh \frac{x_b}{L_{n+}}} \right\}. \tag{7.10}$$

If the highly doped layer is thin enough, Eq. 7.10 shows that S_{eff} will be reduced by the ratio of doping levels.

- Passivation of front surface with thin oxide coating. The high surface recombination velocity at a free silicon surface tends to create a dead layer, where photogenerated carriers are not collected, at the surface of an unpassivated cell. Oxidising the surface creates a thin layer of the wide band gap insulator, silicon dioxide, which prevents carriers from reaching the surface and hence reduces the effective surface recombination velocity. The interface between silicon and silicon dioxide is much less defective than a free silicon surface. This reduces the loss of carriers in the emitter through surface recombination, and improves the response to blue light.
- Use of point contacts at rear. Since the silicon–metal interface is more defective than the silicon–silicon dioxide interface, rear surface recombination can be reduced by contacting only part of the rear p layer with metal, using 'point' contacts. The rest of the surface can then be passivated with oxide, and the overall surface recombination losses greatly reduced. In order to avoid problems with series resistance, the region of semiconductor close to the point contacts is differentially doped p^+. This innovation followed from the rear point contact solar cell, described below.

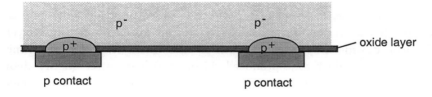

Fig. 7.12. Rear point contacts in p type material.

7.4.6. *Strategies to reduce series resistance*

- Optimisation of the n region doping. Reduced doping improves collection from the n region, giving a better response to blue light. Increased n doping increases V_{bi} and reduces series resistance, although very high n doping is unhelpful for increasing V_{oc} because of Auger recombination and band gap narrowing.
- Differential doping of the area around the contacts. This is achieved by exposing the areas to be contacted to dopant rich gases before deposition of the contacts. For point and grid contacts, the current density through the material close to the contacted area will be high. Doping this volume heavily reduces the losses to series resistance.
- Narrow but deep fingers in front contact, as above. The high aspect ratio reduces surface area blocked by contacts without reducing finger cross sectional area, and the relatively high contact area between fingers and semiconductor reduces the current density at the contact.

7.4.7. *Evolution of silicon solar cell design*

Let's look at how these strategies have been incorporated into the design of silicon solar cells. A summary of crystalline silicon cell performance data is given in Table 7.2.

Black cells

Typical 'black cell' designs (so called because of their almost zero reflectivity) were developed in the early 1980s and exhibited efficiencies of up to 17%. Black cells incorporated the innovation of the surface texturing as well as the features of the basic cell described above.

Passivated emitter cells

Passivated emitter solar cells (PESC) are so called because of the innovation of the passivation of the non-contacted front surface with a thin layer of silicon dioxide. Improvements such as these make it worthwhile using more expensive float zone produced silicon, which is better quality than Czochralski and has a longer diffusion length. The PESC cell was designed at the University of New South Wales and achieved an efficiency of 20% in 1985.

Table 7.2. Record efficiencies for various silicon cells and the best GaAs cell under standard test conditions (Air Mass 1.5 direct, 25°, 1000 W m^{-2}). [Green *et al.*, 2001; Green, 1995.]

DESCRIPTION	WHERE DEVELOPED	AREA (CM^2)	V_{OC} (V)	J_{SC} (MA/CM^2)	FF	EFFICIENCY (%)
monocrystalline Si cells:						
PESC	UNSW	12				20.8
Buried contact	UNSW	10				21.3
Back point contact	Stanford					22.3
PERL	UNSW	4.0	0.696	42.0	83.6	24.4
polycrystalline Si	UNSW/Eurosolare	1.0	0.628	36.2	78.5	19.8
commercial Si		100				13
monocrystalline GaAs	Kopin	3.91	1.022	28.2	87.1	25.1

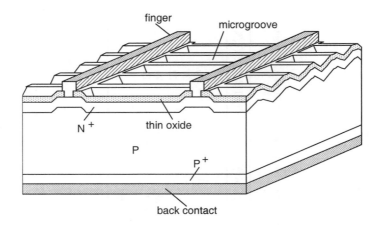

Fig. 7.13.   Passivated emitter solar cell (PESC) [Green, 1995].

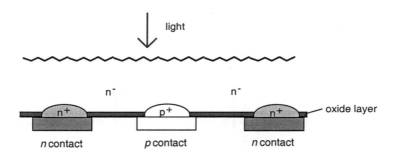

Fig. 7.14.   Schematic cross section of rear point contact cell.

Rear point contact cell

By placing both the n and p contacts on the rear of the cell, this design
eliminates shading losses entirely. This cell was introduced at Stanford in
1992, with an efficiency of 22%. The original design was intended for use
in concentrators [Sinton, 1986]. The cell is made from lightly doped n type
silicon with heavily doped n and p type regions close to point contacts on
the rear surface. The front surface is passivated and textured as usual. The
cell is thin (100 μm) and is intended to operate at high injection levels, so
light trapping is important.

Extremely high purity material is needed, because photogenerated carriers have to *diffuse* to the rear of the cell. Small space charge regions will

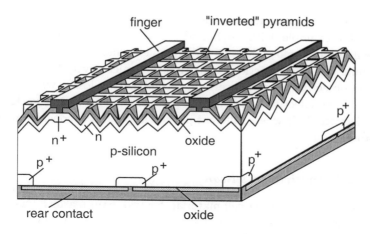

Fig. 7.15. PERL cell [Green, 1995].

develop at the rear of the cell between contacts of opposite polarity rather than at the front. Another difficulty is the risk of shorting out between contacts of opposite polarity on a single surface.

PERL cell

The passivated emitter, rear locally diffused (PERL) solar cell was developed at UNSW, with an efficiency of 24% in 1994 [Zhao, 1994]. This design exploits the advantage of point contacts in reducing recombination at the rear surface. It has the following features:

- Rear point contacts reduce the area of the semiconductor-metal interface, where recombination is high, so that most of the rear surface may be contacted with oxide.
- Grooved front contacts as with the passivated emitter solar cell.
- Differential heavy doping of n layer near contacts
- Surface texturing using inverted pyramids

7.4.8. *Future directions in silicon cell design*

The performance of silicon solar cells is now fairly close to the theoretical maximum of 29%. Continuing refinements to the design, mainly aimed at reducing shading and series resistance losses, may increase efficiencies of lab cells to 26% or 27% in AM1.5. The main challenges are now in improving cell production techniques in order to mass-produce efficient cells more cheaply. For example, with the buried contact cell, efforts have focused on producing grooves more cheaply, for example by mechanical etching.

A quite different direction is the thin film microcrystalline silicon cell. Here the objective is to reduce bulk recombination losses without losing absorption and effective light trapping is required. This design works in the 'high injection' limit where different physics applies. It is considered briefly in Chapter 8.

7.4.9. *Alternatives to silicon*

Silicon is not an ideal solar cell material, for two main reasons. One is that its band gap (1.1 eV) is smaller than the optimum (1.4 eV) for terrestrial solar energy conversion. The other is that since its absorption coefficient is low, a relatively thick layer of silicon is needed (in conventional silicon designs) to absorb sunlight effectively. The significant requirement for high purity silicon increases the cost, as well as the weight, of the cell. Another consideration is the temperature dependence of efficiency which makes silicon less suitable for applications under concentrated light and in space. Therefore a number of alternative semiconductor materials have been developed for photovoltaic applications. Cell designs aimed at reducing costs by using less or less pure semiconductor material are considered in Chapter 8. In the next section we will briefly consider materials, from the III–V group of compound semiconductors, with more favourable materials properties for high efficiency single crystal cells.

7.5. III–V Semiconductor Material Properties

7.5.1. *III–V semiconductor band structure and optical absorption*

A III–V semiconductor is an alloy containing equal numbers of atoms from groups III and V in the periodic table. The group III atom contributes three valence electrons to bonding and the group V element contributes five. For many compounds the atoms arrange themselves into the *zincblende* crystal structure — two interlocking face centred cubic lattices — where each atom forms four bonds to neighbouring atoms of the other type. All of the valence electrons are used up in bonding, and, as in the case of silicon, the lowest energy configuration for the crystal is a semiconductor where

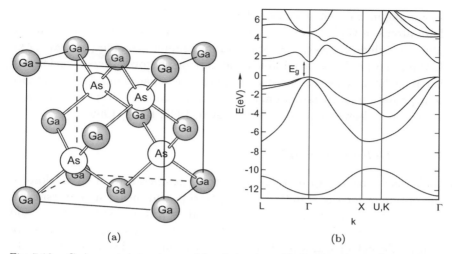

(a)   (b)

Fig. 7.16. GaAs crystal structure and band structure. Notice that the conduction band maximum occurs at the same point as the valence band minimum (the Γ point, where $k = 0$, giving a direct band gap).

there is a band gap between the (normally empty) conduction band and the (normally filled) valence band.

Relative to silicon, III–V's have several advantages as electronic materials. One is the possibility of varying the crystal composition by replacing some of the group III atoms with another group III element in order to vary the band gap in a controlled way. Another is that for many compositions these materials are *direct* gap semiconductors, and so are much more effective optical absorbers (see Fig. 7.3).

III–Vs have been widely developed for applications in optoelectronics. They are grown by a number of epitaxial techniques such as liquid phase epitaxy (LPE), molecular beam epitaxy (MBE), metal-organic chemical vapour deposition (MOCVD) and metal-organic vapour phase epitaxy (MOVPE) which allow minute control of the composition and layer thickness. Epitaxial growth has been developed in particular so that heterostructures — layered structures of materials of different band gap which enable spatial confinement of carriers for applications such as lasers — may be fabricated. The best understood and most widely used of these III–V semiconductors is *gallium arsenide* (GaAs). It is also the most suitable for solar energy conversion. Other relevant materials are the binary alloys indium phosphide (InP), gallium antimonide (GaSb) and ternary alloys such as aluminium gallium arsenide ($Al_xGa_{1-x}As$), where a fraction x of the gallium atoms in GaAs have been replaced by aluminium atoms, indium gallium phosphide ($In_xGa_{1-x}P$) and indium gallium arsenide ($In_xGa_{1-x}As$). Indium phosphide has a suitable band gap for photovoltaic conversion and is particularly attractive for space applications on account of its resistance to degradation under radiation, or 'radiation hardness'. Properties of GaAs are covered elsewhere [Lush, 1986; Adachi, 1992; Bube, 1998].

7.5.2. *Gallium arsenide*

GaAs has a band gap of 1.42 eV at room temperature. This is close to the optimum for the standard solar spectrum and means a conversion efficiency of 31% is theoretically possible. It is a direct gap material and absorbs strongly above its band gap. Over visible wavelengths the absorption coefficient of GaAs is about ten times that of silicon (see Fig. 7.3), and only a few μm rather than hundreds of μm are needed for the active region of the solar cell. This most important for space applications where the priority is to reduce the cell weight.

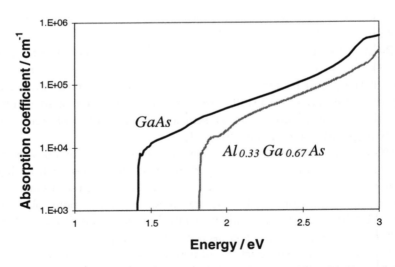

Fig. 7.17.   Absorption coefficient of GaAs and $Al_{0.33}Ga_{0.67}As$. Note the strong absorption at energies just above the band gap, and the similar shape of the spectra. As the aluminium fraction x is increased, the $Al_xGa_{1-x}As$ band edge moves to higher energies. At compositions above $x = 0.4$ the band gap becomes indirect and $Al_xGa_{1-x}As$ is too poor an absorber to be useful for photogeneration. However, high aluminium content material is used as a 'window', much as oxide is used in silicon cells.

GaAs enjoys other advantages for solar cell applications. One is a better temperature coefficient than silicon. Solar cell efficiency tends to decrease as temperature increases, because of increasing carrier recombination and a decreasing band gap. The first effect is more pronounced in silicon where recombination depends upon the availability of phonons, which increases with temperature. This means that GaAs performs better in situations where the cell operates at high temperature, under concentration and in space.

Another factor relevant to space is the radiation resistance of the material. Exposure to extraterrestrial radiation degrades the efficiency of a solar cell over its lifetime. GaAs has a better radiation hardness than silicon, although indium phosphide is better still.

High purity GaAs is much more costly than pure silicon, and GaAs cells are some 5 to 10 times as expensive in spite of reduced cell thickness. However, GaAs production technology is still maturing, and production costs are decreasing as the scale of manufacture of GaAs based devices for applications in optoelectronics expands.

These factors mean that GaAs cells have been developed primarily for use in space. The most likely terrestrial application is for power generation under concentrated light. (Concentration is discussed in Chapter 9.) Although most of the design features mentioned below are general, some were developed with concentration in mind.

7.5.3. *Doping*

III–V semiconductors can be doped by replacing one of the elements with one of different valence. In GaAs, n type doping may be achieved by introducing controlled amounts of silicon during growth. The tetravalent silicon atoms normally replace some of the trivalent gallium atoms in the lattice, introducing a donor state associated with the extra valence electron. Tin, which is also tetravalent, has also been used as a donor impurity. For p type doping, carbon is the most commonly used impurity. Carbon, like silicon, is tetravalent but under certain growth conditions it prefers to substitute for *arsenic* atoms in the lattice, and so introduces a deficiency of valence electrons and hence acceptor states. Alternatively the group II element beryllium can be used, which introduces an acceptor state by substituting for gallium atoms. Dopants may be introduced by diffusion or directly during growth by epitaxial techniques.

7.5.4.   *Recombination*

Radiative recombination is faster in GaAs than in silicon, and may dominate in very pure material. The direct band gap means that direct transitions from conduction to valence band are more likely and the radiative lifetime is a factor of 10^4 shorter. For example, at acceptor doping levels of 10^{16} cm^{-3}, τ_{rad} is around 1 μs in pure p type GaAs, compared to around 10 ms in silicon [Lush, 1986; Green, 1995]. The large difference is due to the difference in the absorption coefficients at energies just above the respective absorption edges. We know from Sec. 4.5.1 that the recombination rate is proportional to the value of α within k_BT of the absorption edge.

Auger recombination, on the other hand, is much slower than in silicon due to the much smaller intrinsic carrier concentration ($n_i = 2 \times 10^6$ cm^{-3} in GaAs, compared to 1×10^{10} cm^{-3} in silicon) and is negligible in photovoltaic devices.

In practical materials, nonradiative Shockley Read Hall recombination through defect states dominates. SRH lifetimes depend on the nature and concentration of impurities and the growth conditions. Electron lifetime in good quality p type GaAs varies from around 1 μs at low doping levels to approximately $(10^{19}/N_a)$ ns at higher doping. For holes in n type GaAs τ_p varies from 10–100 ns at low doping to a few ns at $N_d > 10^{18}$ cm^{-3}. τ_p is up to one order of magnitude shorter than τ_n in p type GaAs doped to the same level.

In p–n devices, SRH recombination in the space charge region dominates performance. This is because the SRH recombination rate is greatest in this region where n and p are similar, and because the carrier densities in the junction region of a GaAs device are high because of the high absorption. The recombination rate therefore involves both carrier types, and is not simply proportional to the minority carrier density, as in silicon, nor is it possible to resolve the lifetime into factors due to different processes, as in Eq. 7.8.

The effect of space charge region recombination is evident in the dark current characteristics of GaAs solar cells. The dark current tends to vary with the ideality factor of 2 characteristic of SRH recombination at the junction, $J_{dark} \sim e^{qV/2k_BT}$, at low positive biases. In silicon cells, for comparison, where the dark current is dominated by recombination in the neutral regions, the ideality factor is close to 1. At higher forward biases in good quality GaAs cells J_{dark} begins to vary like e^{qV/k_BT}, as radiative recombination begins to dominate.

Table 7.3.   Typical GaAs material and cell parameters, used in calculations in Fig. 7.19.

	EMITTER (P TYPE)	BASE (N TYPE)
Thickness/μm	0.5	4
Doping/cm^{-3}	1×10^{18}	1×10^{17}
Minority carrier diffusion constant/cm^2 s^{-1}	25	10
Minority carrier lifetime/s	1×10^{-9}	1^{-8}
Minority carrier diffusion length/μm	1.5	3
Surface recombination velocity/cm s^{-1}	1000000	1000000
Reflectivity	0.05	
Absorption: at 500 nm/cm^{-1}	110000	
at 800 nm/cm^{-1}	13000	

Like bulk recombination, surface recombination is higher in GaAs than in silicon. At untreated surfaces the minority carrier surface recombination is of the order of 10^6 cm s^{-1} [Hovel, 1975] but this can be reduced to under 10^4 cm s^{-1} using window layers, discussed below [Bube, 1998].

7.5.5.   *Carrier transport*

As in silicon, electrons have a higher mobility than holes in GaAs. Minority electrons have a mobility of around 5000 cm^2 V^{-1} s^{-1} at low doping which falls to around 1000 cm^2 V^{-1} s^{-1} at doping levels of 10^{18} cm^{-3}. The minority hole mobility varies from 300–400 cm^2 V^{-1} s^{-1} at low doping levels to less than 100 cm^2 V^{-1} s^{-1} at doping levels of more than 10^{18} cm^{-3}. Together with minority carrier lifetimes, these lead to diffusion lengths of a few μm in moderately doped GaAs, longer for p than n type. In ternary alloys, the higher defect density leads to much faster recombination. For example, in n doped $Al_xGa_{1-x}As$, L_p may be less than 0.1 μm and lifetimes may be a few tens of ps.

The doping dependent resistivity of p type GaAs is around $(10^{15}/N_a)$ Ohm cm. It is about an order of magnitude greater for n type material doped to the same level.

7.5.6.   *Reflectivity*

GaAs has a refractive index of 3.3 and a natural reflectivity of 30–40% over visible wavelengths. Suitable anti-reflection coatings can be made from silicon nitride.

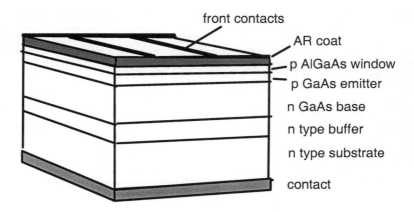

Fig. 7.18.   Layer structure of a windowed monocrystalline GaAs solar cell.

7.6.   GaAs Solar Cell Design

7.6.1.   *Basic GaAs solar cell*

In GaAs, because diffusion lengths greater than the absorption depth can be achieved for either doping type, cells can be prepared as either *p–n* or *n–p* designs. In either case the emitter should be as thin as possible without increasing series resistance too much. For the *p–n* design, a 0.5 μm emitter doped to 10^{18} cm^{-3} is typical; for the *n–p* design, the emitter can be as thin as 0.2 μm because of the higher n type conductivity. In practice p^{+}–n designs seem to perform better than n^{+}–p designs. The base is much shorter than in silicon cells, typically 2–4 μm and comparable with the diffusion length.

7.6.2.   *Optimisation of GaAs solar cell design*

Figure 7.19 illustrates absorption and recombination profile in a typical GaAs *p–n* junction cell, calculated as Fig. 7.8 and using the materials parameters in Table 7.3. The graphs show that for GaAs the influences on recombination are rather different. We can make the following observations:

- absorption of light is good at all wavelengths
- front surface recombination is important for long wavelengths
- recombination in the junction region is dominant

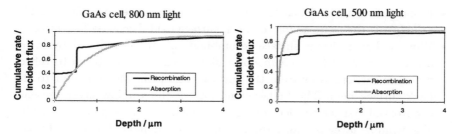

Fig. 7.19. Simulated cumulative absorption and recombination rates for a GaAs solar cell with the parameters in Table 7.3 under monochromatic light at two different wavelengths and at a bias close to open circuit.

- bulk recombination is unimportant relative to junction and surface recombination
- rear surface recombination is negligible, because of the high absorption

Therefore, the objectives in optimising GaAs cell design should be to

- Minimise front surface recombination
- Minimise junction recombination
- Minimise series resistance

We will add one more very practical objective:

- Minimise substrate cost.

This arises because the GaAs layers are extremely thin, and must be grown on a substrate for mechanical stability, yet depositing GaAs cells on GaAs substrates is prohibitively expensive.

Strategies to address these issues are discussed below.

7.6.3. *Strategies to reduce front surface recombination*

Because of the high absorption coefficient, more carriers are generated in the emitter in a GaAs than a silicon cell. This means the contribution of the emitter to the photocurrent is not negligible and also that surface and bulk recombination in this region are important.

- Front surface fields. A heavily doped layer is introduced by diffusion. The principle is the same as for the back surface field discussed for silicon.
- Window layers. Front surface recombination can be reduced by introducing a front surface *window* of a higher band gap material to reflect minority carriers away from the surface. The principle is similar to the back surface field. The higher band gap of the window layer presents a potential barrier to electrons generated in the p region (Fig. 7.20). The window layer is transparent to most visible light, but the interface between the window and the bulk GaAs is much less defective than untreated GaAs. The consequence is that the effective front surface velocity is greatly reduced. An analysis similar to that given above for back surface field shows that, if interface recombination is negligible, the effective surface velocity at a window layer is given by

$$S_{\text{eff}} = \frac{D_{\text{n}+}}{L_{\text{n}+}} e^{\Delta E_{\text{g}}/kT} \left\{ \frac{\frac{S_{\text{n}+} L_{\text{n}+}}{D_{\text{n}+}} \cosh \frac{x_{\text{b}}}{L_{\text{n}+}} + \sinh \frac{x_{\text{b}}}{L_{\text{n}+}}}{\frac{S_{\text{n}+} L_{\text{n}+}}{D_{\text{n}+}} \sinh \frac{x_{\text{b}}}{L_{\text{n}+}} + \cosh \frac{x_{\text{b}}}{L_{\text{n}+}}} \right\} \qquad (7.11)$$

where ΔE_{g} is the difference in band gaps between GaAs and the window and $n+$ denotes the window layer. Typical window materials are $Al_x Ga_{1-x} As$ and $In_x Ga_{1-x} P$, both of which are lattice matched to GaAs and so result in good quality interfaces with a low density of interface states [Hovel, 1975].
- Heterojunctions and graded emitters. Other options are to fabricate the whole of the emitter from a wider gap material than the base, producing a *p–n heterojunction* cell, where the *p–n* junction actually occurs at the junction between two materials of different composition. The advantage is that the emitter still absorbs short wavelengths but recombination is suppressed, so improving response to blue light. The wider gap emitter also offers the possibility of increases in V_{oc} (though in practice these are negligible) and flexibility to maintain high doping in emitter without the recombination losses which would result in a homojunction cell. However, the interface is likely to introduce defects, and since recombination is already highest in the depleted region where n and p are similar, there is a danger that defects at the heterojunction will assist recombination. A compromise is to locate the high band gap — low band gap interface away from the depleted region in the neutral p layer.

Another idea is to grade the composition of the p layer, from GaAs near to the junction, to a high band gap alloy at the front surface. The compositional grade introduces an electric field which assists electron migration to

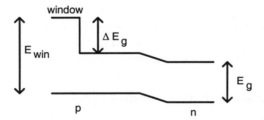

Fig. 7.20.   Band diagram of a *p–n* GaAs cell with a *p* type AlGaAs window of band gap E_{win}.

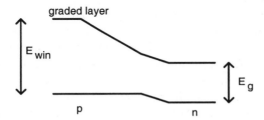

Fig. 7.21.   Band diagram of a GaAs cell with an $Al_xGa_{1-x}As$ emitter of graded composition. The compositional grade provides an effective electric field which accelerates electrons towards the junction region and assists charge separation.

the junction, as shown in Fig. 7.21. This can be done in GaAs cells using $Al_xGa_{1-x}As$ of varying aluminium fraction and has resulted in enhanced photocurrent through improved minority carrier collection. The wider gap acts like a window layer, shifting the absorption of longer wavelength photons towards the junction where they are more likely to be utilised. These strategies would certainly be useful for silicon cells, but with silicon there is no convenient wider band gap semiconductor with a similar lattice constant which could be used.

7.6.4.   *Strategies to reduce series resistance*

Series resistance is an issue with all solar cell designs because of the need to expose as much of the front surface as possible to the light, yet make electrical contact to it to collect the current. In GaAs concentrator cells it is a particularly important issue. For a cell to operate at optimal efficiency under 100 suns, the sheet resistance of the cell should be less that 10^{-3} Ohm-cm^2. GaAs cells for use in concentration are therefore designed to have minimal resistance at high injection levels, and are not necessarily the optimum design for one sun. This means special grid patterns with high metallisation, and very high emitter doping, which can be achieved with the help of a window layer and a front surface field.

7.6.5. *Strategies to reduce substrate cost*

Because the GaAs cell is so thin, it must be grown on a substrate, and it is better to choose a substrate of the same lattice constant as the cell material in order not to introduce crystal defects at the rear interface. GaAs substrates are ideal, but prohibitively expensive, and for large scale production germanium, which has a similar lattice constant, is used. Ge has the same lattice constant as GaAs, but is rather rare and in the end the cost of GaAs on Ge solar cells is likely to be limited by the cost of Ge. Growth of GaAs on Si substrates suffers from the difference in lattice constants, and poor material quality inevitably results.

One solution is to re-use the substrate. Techniques have been developed in which epitaxially grown GaAs wafers are removed from a GaAs substrate by etching or cleaving along an intermediate plane. Cells produced this way are brittle and still require mechanical support. Another approach to cost reductions is to use *polycrystalline* GaAs, in which case lattice matching with a substrate is less important. 20% efficient poly-GaAs cells have been grown on Germanium substrates. A long term objective would be to deposit poly-GaAs cells on silicon, metal or glass.

7.7. Summary

The most efficient single junction photovoltaic cells are based on p–n junctions in monocrystalline semiconductors. Crystalline silicon is the most important photovoltaic material and is widely used for terrestrial applications. Crystalline GaAs is more suitable for applications in space and under concentration. Both materials have suitable band gaps leading to a high theoretical efficiency.

The objectives in solar cell design are to maximise light absorption, charge separation and charge transport, and to maximise the photovoltage. The design factors which may be varied in a p–n junction solar cell are the layer thicknesses, the doping levels, the surface treatments and the front surface contact pattern. High light absorption is achieved by using a high optical depth, minimising shading and reflection, and by using light trapping structures. Charge separation is maximised by using a good quality junction with a high built-in bias. Efficient charge transport requires good quality crystal with low bulk and surface recombination and good majority carrier conductivity. For any material the ideal p–n structure has a thin heavily doped emitter on top of a thicker, heavily doped base. The remaining priorities depend upon the material used.

Silicon has a low absorption coefficient and most photogeneration takes place in the base. p type silicon is used as the base because the minority carrier transport characteristics are better than in n type material. The base should be several hundred μm thick for good light absorption, while the n type emitter is typically less than 0.5 micron thick. The base is lightly doped so that electron diffusion lengths are long enough (several hundred μm) for good charge collection. Because of the low absorption, the priorities in silicon solar cell design are to minimise light reflection and shading, and to reduce recombination at the rear surface. Reflection is reduced by texturing the front surface with a pattern of micron sized pyramids and by adding an anti-reflection coat. Shading is reduced by using thin contact fingers, and by burying the contacts into the emitter, while the series resistance which results from thin contacts can be reduced by differential doping of the semiconductor near to the contacts. Surface recombination is reduced by passivating the surface with a layer of silicon dioxide, and by the use of point contacts at the rear. The best monocrystalline silicon cell is 24% efficient in AM1.5.

GaAs is a strong absorber and devices only a few microns thick absorb most of the light. Both n–p and p–n structures may be used, with typical emitter thicknesses of less than 0.5 μm and base thicknesses of 2–4 μm. Significant photogeneration takes place in the emitter and junction as well as the base. This means that recombination at the front surface and in the junction is important. Front surface recombination can be reduced by building in doped window layers from a wider band gap semiconductor. The window separates the active region from the metallic contact, and helps to drive minority carriers towards the junction. Junction recombination is minimised mainly by control of the crystal quality and doping levels. Reflection is minimised using anti-reflection coats; texturing is not suitable because of the larger size of light trapping structures compared to the device thickness, and the sensitivity of the front surface quality. As in silicon, the front surface contact pattern should be designed to minimise shading without enhancing series resistance. The best single cell efficiency, of 25.1%, has been achieved in GaAs.

References

S. Adachi, *Physical Properties of III-V Semiconductor Compounds* (Chichester: Wiley, 1992).

R. Bube, *Photovoltaic Materials* (Imperial College Press, 1998).

M.A. Green, *Silicon Solar Cells: Advanced Principles and Practice* (Sydney: Centre for Photovoltaic Engineering, 1995).

M.A. Green, K. Emery, D.L. King, S. Igari and W. Warta, "Solar cell efficiency tables", *Progress in Photovoltaics* **9**, 287–293 (2001).

H.J. Hovel, "Solar cells", in *Semiconductors and Semimetals* **11**, eds. R.K. Willardson and A.C. Beer (1975).

G.B. Lush, in *Properties of Gallium Arsenide* (INSPEC, 1986).

M. Shur, *Physics of Semiconductor Devices* (Prentice Hall International, 1990).

R. Sinton *et al.*, "27.5% silicon concentrator solar cells", *Electron Device Letts.* **EDL-7**, 567 (1986).

R.J. van Overstraeten and R. P. Mertens, *Physics Technology and Use of Photovoltaics* (Bristol: Adam Hilger, 1986).

J. Zhao and M.A. Green, "23.5% efficient silicon solar cell", *Prog. Photovoltaics* **2**, 227–230 (1994).

Chapter 8

Thin Film Solar Cells

8.1. Introduction

In Chapter 6 we covered the theory of ideal p–n junction solar cells, and in Chapter 7 applied it to monocrystalline silicon and GaAs devices. We have seen that, provided that minority carrier diffusion lengths exceed typical absorption depths, p–n junctions make efficient photoconverters with a high collection efficiency, where recombination at the surfaces is the dominant loss process. However, single crystals are expensive to produce and so there is a great deal of interest in finding photovoltaic materials of less demanding material quality which can be grown more cheaply. A number of materials have been identified of which the best developed at present are amorphous silicon (a-Si), polycrystalline cadmium telluride (CdTe), polycrystalline copper indium diselenide (CuInSe$_2$) and microcrystalline thin film silicon (p-Si). These 'thin film' materials are usually produced by physical or chemical deposition techniques which can be applied to large areas and fast throughput. Note that the term 'thin film' refers more to solar cell technologies with mass-production possibilities rather than the film thickness: GaAs p–n junction cells, with an active layer a few μm thick, are thin, but do not belong to this class.

Polycrystalline and amorphous semiconductors contain intrinsic defects which increase the density of traps and recombination centres. For solar cells, this has the consequence that:

- Diffusion lengths are shorter, so the material needs to be a strong optical absorber. Alternatively, multiple junctions must be used to make the device optically thick. In the case of very short diffusion lengths, it may be necessary to use extended built-in electric fields to aid carrier collection. This is the case in amorphous silicon, where p–i–n structures are preferred.

- Losses in the layers close to the front surface are greater, so it is advantageous to replace the emitter with a wider band gap window material.
- The presence of defect states in the band gap can make the materials difficult to dope, and can limit the built-in bias available from a junction through Fermi level pinning.
- The presence of grain boundaries and other intrinsic defects increases the resistivity of the films particularly at low doping densities, and makes the conductivity dependent on carrier density, so influencing the electrical characteristics of devices.
- The presence of defects similarly means that minority carrier lifetime and diffusion constant are carrier density dependent.

What this means for the calculation of solar cell current-voltage characteristics is that (i) the model of the p–n junction developed in previous chapters must be modified for p–i–n structures and p–n heterojunctions and (ii) the solution of the transport equations becomes more complicated, because the parameters such as mobility, lifetime and diffusion constant are functions of carrier density. The differential equations of the form

$$\frac{\partial n}{\partial t} = \frac{1}{q}\nabla \cdot J_{\mathrm{n}} + G_{\mathrm{n}} - U_{\mathrm{n}}$$

which result from electron continuity, are no longer linear in n. In the course of this chapter we will show how the simple model of the p–n junction can be adapted for other device structures, and indicate how current generation and recombination are influenced by the presence of defects.

The chapter is organised as follows: first we discuss the general features of thin film photovoltaic materials. Then we focus on amorphous silicon, reviewing the materials properties, the consequences for charge transport and photocurrent generation, and the design of amorphous silicon solar cells. At this stage the analysis of the p–i–n junction is presented. Then we move on to polycrystalline thin film materials. In this section a simple model of a grain boundary is presented. Finally the design of CdTe, CuInSe$_2$ and thin film silicon solar cells is discussed.

Note that this chapter is not intended as a comprehensive review of the materials science and technology of thin film solar cells, but rather to highlight certain aspects of the device physics. Reviews of the state of thin film photovoltaics for the different materials are found in Bube [Bube, 1998] and Archer and Hill [Archer, 2001].

8.2. Thin Film Photovoltaic Materials

8.2.1. *Requirements for suitable materials*

Good thin film materials should be low cost, non-toxic, robust and stable. They should absorb light more strongly than silicon. Higher absorption reduces the cell thickness and so relaxes the requirement for long minority-carrier diffusion lengths, allowing less pure polycrystalline or amorphous materials to be used. Figure 8.1 compares the absorption coefficients for several photovoltaic materials and the maximum photon current which can be generated in a thin film as a function of its thickness. Notice how weakly crystalline silicon absorbs, in comparison with the other materials. Suitable materials should transport charge efficiently, and should be readily doped. Materials are particularly attractive if they can be deposited in such a way that arrays of interconnected cells can be produced at once (Fig. 8.2). This greatly reduces the module cost.

Of the elemental semiconductors, only silicon has a suitable band gap for photovoltaic energy conversion. Compound semiconductors greatly extend the range of available materials and of these a number of II–VI binary compounds and I–III–VI ternary compounds have been used for thin film photovoltaics. Many of these are direct band gap semiconductors with high optical absorption relative to silicon. The I–III–VI compounds (or chalcogenides) are analogous to II–VI's where the group II element has been replaced by a group I and a group III species. At present the leading compound semiconductors for thin film photovoltaics are the II–VI semiconductor, CdTe, and the chalcogenide alloys, $CuInGaSe_2$ and $CuInSe_2$. Other new materials are continually being investigated, including other II–VI and I–III–VI compounds, amorphous carbon and nanocrystalline silicon. Molecular electronic materials form a new class of thin-film photovoltaic materials, but rely on different physics, and they will not be discussed here.

8.3. Amorphous Silicon

8.3.1. *Materials properties*

Amorphous silicon (a-Si) is the best developed thin film material and has been in commercial production since 1980, initially for use in hand held calculators.

As a material for photovoltaics, it has the advantages of relatively cheap, low temperature ($< 300°C$) deposition and the possibility of growing on a

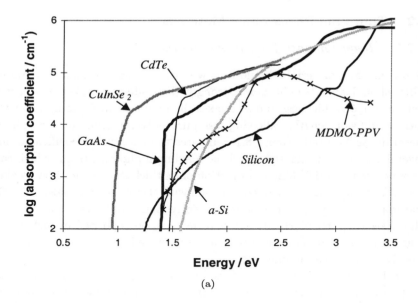

(a)

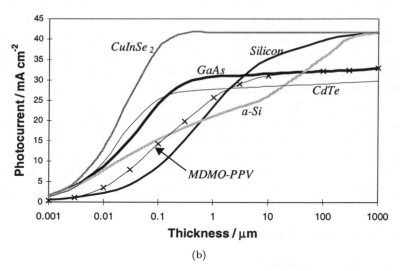

(b)

Fig. 8.1.   (a) Absorption coefficients of a number of different photovoltaic materials.
(A derivative of the organic semiconductor polyphenylene vinylene (MDMO-PPV) is
included for comparison.) [Mitchell, 1977; Tuttle, 1987; Fritzsche, 1985]; (b) Maximum
photon current available from each material under AM1.5 illumination, as a function of
film thickness, assuming perfect collection of all photogenerated charges. The saturation
photocurrent is a function of band gap. The maximum photocurrent supplied by the
AM1.5 spectrum is around 49 mAcm^{-2}.

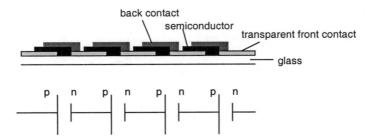

Fig. 8.2. Integral interconnections in a thin film cell. To form integrally interconnected modules the layers are scribed between stages first to separate individual cells and then to connect them in series. This removes the need for separate contacting and connecting of cells, which is costly in conventional module designs.

variety of substrates, including glass, metal and plastic, with diverse commercial applications. The amorphous nature has several important consequences for photovoltaics. Absorption of visible light is better than for crystalline silicon, but doping and charge transport are more difficult. The availability of alloys with different band gap enable the design of heterostructure and tandem devices.

8.3.2. *Defects in amorphous material*

In amorphous materials, the lattice contains a range of bond lengths and orientations, as well as unsatisfied 'dangling' bonds. Although the nearest neighbours of any atom are co-ordinated almost exactly as in the crystalline material, the combined effect of small bond distortions means that there is virtually no correlation between an atom and its more distant neighbours. The long range order of the crystal is gone. In a-Si, Si atoms are arranged in an approximate tetrahedral lattice but with a variation of up to $10°$ in the bond angles. As mentioned in Chapter 4, this loss of order means that the selection rules for photon absorption are relaxed, and a semiconductor which is indirect in its crystalline form behaves like a direct gap material in its amorphous form. This increased absorption is one reason why amorphous silicon is of interest for photovoltaics.

The variation in Si–Si distance and orientation gives rise to a spreading in the electron energy levels, relative to the perfect crystal. This spreading appears as a tail in the density of states at the top of the valence band and the bottom of the conduction band, known as an *Urbach* tail. *Dangling bond* states are due to Si atoms which are co-ordinated only to three

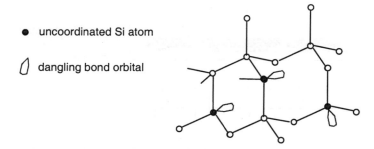

Fig. 8.3.   Dangling bonds in amorphous silicon in which some silicon atoms are bonded to only three rather than four neighbours, leaving an unused valence orbital. This defect is known as a dangling bond, and it may be neutral, positively or negatively charged.

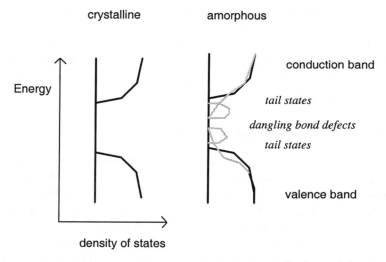

Fig. 8.4.   Schematic density of states in amorphous material, compared to the crystalline form.

neighbouring Si atoms, leaving one valence orbital which is not involved in bonding. The dangling bond may be positively charged (D^+), neutral (D^0) or negatively charged (D^-). An excess of D^- states gives rise to n type a-Si while excess D^+ gives rise to p-type material. More defects can be created by irradiating the material, by heating and sudden cooling, or by extrinsic doping. The dangling bond states give rise to energy levels deep in the band gap (Fig. 8.4).

The defects in amorphous material differ from those in polycrystalline materials in that they occur uniformly throughout the material and not only at grain boundaries.

8.3.3. *Absorption*

The loss of crystal order means that the absorption coefficient is higher than in crystalline silicon. In crystalline silicon the band gap is indirect for most visible wavelengths and so absorption of a photon requires the simultaneous absorption of a phonon to conserve crystal momentum. Absorption thus is limited by the availability of phonons. In amorphous silicon there is no well defined E–k relationship and no requirement to conserve crystal momentum. Absorption depends simply upon the availability of photons and the density of states in valence and conduction bands, much as it does in direct gap crystals. The absorption coefficient is about an order of magnitude greater than in crystalline silicon (c-Si) at visible wavelengths (see Fig. 8.1).

Passivation of a-Si with hydrogen (discussed below) increases both the absorption coefficient and the band gap. Material used for solar cells typically has a band gap of 1.7 eV. This is higher than the optimum for solar energy conversion, but material with lower concentrations of hydrogen is unusable due to poor doping and transport properties.

8.3.4. *Doping*

In unpassivated a-Si the defect density is so high ($> 10^{16}$ cm^{-3}) that the material cannot normally be doped. The extra carriers which would be introduced into the conduction or valence band by donor or acceptor impurities are captured by dangling bond defects. However, the background density of defect states may be reduced by saturating the dangling bonds with atomic hydrogen (Fig. 8.5). Hydrogen forms a bond with the unpaired electron in a neutral (D_0) defect and removes the capacity of that defect to trap an electron or a hole. Passivation with 5–10% hydrogen reduces the density of dangling bonds to around 10^{15} cm^{-3}, and produces material from which workable p–n junctions can be made.

The material may be doped n or p type but the doping efficiency is low. For example, doping with phosphorus introduces a density N_d of neutral P atoms which in crystalline silicon would normally ionise to add one electron

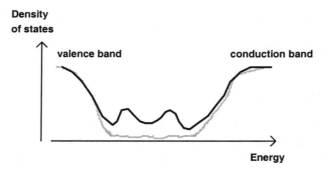

Fig. 8.5.   Schematic of density of states before (black) and after (grey) passivation in amorphous silicon. Unpassivated dangling bond defects are responsible for peaks in the density of states deep in the band gap. Before passivation the defect density is so high that the material cannot be doped.

to the conduction band:

$$P^0 \rightarrow P^+ + e^-.$$

In amorphous silicon, however, the addition of an electron upsets the equilibrium between neutral and negatively charged dangling bonds, and drives the following reaction to the right:

$$e^- + D^0 \Leftrightarrow D^-$$

so that the density of D^- states increases. This decreases the minority carrier lifetime since D^- states are recombination centres for holes. It can be shown that the density of D^- increases like $\sqrt{N_d}$ and therefore that the density of majority carriers increases more slowly than N_d. The effect on the Fermi level is to pull the Fermi level away from the donor or acceptor level towards the defect levels in the band gap, as shown in Fig. 8.6. When the density of dangling bonds is very high, the Fermi level is *pinned* amongst the defect states. Low doping efficiency means that the majority carrier activation energy — which is the difference between Fermi level and band edge — is high. (In crystalline material this should be equal to the difference between the impurity level and the band edge.) In p-type a-Si the activation energy is around 0.4 eV. These large activation energies limit the size of the built-in bias which can be achieved at a p–n or p–i–n junction, increasing the difference between the built-in bias and the band gap.

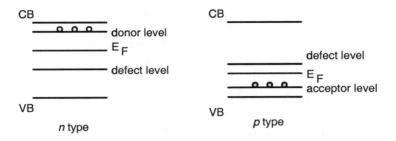

Fig. 8.6. Fermi levels in p and n type amorphous material.

8.3.5. *Transport*

The defect states which remain after hydrogen passivation act both as charge traps and recombination centres and dominate charge transport in a-Si.

The distribution of tail states below the conduction and valence band edges act as traps for mobile carriers. Charge carriers in these states move by a sequence of thermal activation and retrapping events. This distribution in energies leads to a distribution in the time constants for any transient process, and a strong dependence upon the occupation of the states, and hence carrier density. Consequently the usual transport parameters of mobility, lifetime, and diffusion constant are density dependent and hard to determine. Transport characteristics can be modelled by including transitions to and from the trap states. The density of tail states in the conduction band, say, is often modelled as an exponential of the form

$$g(E) = \frac{N_t}{k_B T_0} \exp\left(\frac{E - E_c}{k_B T_0}\right)$$

where N_t is the volume density of tail states, and T_0 is a characteristic 'temperature' which describes the depth of the tail. Such approaches successfully reproduce the features of transport measurements in a-Si. (For a review, see Tiedje [Tiedje, 1984].) Charge transport in such a defective medium is sometimes called *dispersive* transport.

The dangling bond states which remain after passivation act as recombination centres and can be treated most simply by including capture of electrons and holes by two discrete levels in the band gap representing the unoccupied and singly occupied state of the defect (Fig. 8.7). [*e.g.* Spear, 1976]. Rate equations can be constructed for the capture and

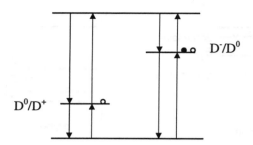

Fig. 8.7. A trivalent defect in amorphous material. The upper level represents the D^-/D^0 state (a singly occupied defect as seen by the electron) and the lower level represents D^0/D^+ (unoccupied defect).

emission of carriers from that level using detailed balance and Fermi Dirac statistics.

In undoped hydrogenated amorphous silicon (a-Si:H) the minority carrier lifetime is typically 10–20 μs and the diffusion length around 0.1 μm, although both are carrier density dependent. Electron mobility is rather better than hole mobility (10^{-6} to 10^{-4} m^2 V^{-1}s^{-1} cf. 10^{-7} to 10^{-6} m^2 V^{-1}s^{-1} for holes), probably due to an asymmetric trap distribution. With doping the defect density increases and diffusion lengths are much reduced. This means that carrier collection in a p–n junction would be extremely poor, and consequently p–i–n structures are used.

8.3.6. *Stability*

Amorphous silicon suffers from light-induced degradation known as the Staebler Wronski effect. The defect density in a-Si:H increases with light exposure, over a time scale of months, to cause an increase in the recombination current and reduction in efficiency. It is believed that light energy breaks some Si-H bonds to increase the density of dangling bonds. The system is excited into in a higher energy configuration with more active defects. Annealing at a few hundred degrees Centigrade allows the structure to relax, and the dangling bonds to be resaturated. For this reason a-Si solar cells may perform rather better in high temperature environments.

In a typical a-Si solar cell the efficiency is reduced by up to 30% in the first six months as a result of the Staebler Wronski effect, and the fill factor falls from over 0.7 to about 0.6. This light induced degradation is the major disadvantage of a-Si as a photovoltaic material.

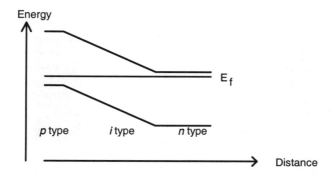

Fig. 8.8.   Band profile of *p–i–n* junction.

8.3.7.   *Related alloys*

a-Si:H may be alloyed with carbon (C) or with germanium (Ge) to produce compound amorphous materials of wider (a-SiC:H) or narrower (a-SiGe:H) band gap. These allow the design of multi-junction, heteroface or graded cells. Materials properties are slightly worse in the alloys; for instance a higher density of valence band tail states in a-Ge.

8.4.   Amorphous Silicon Solar Cell Design

8.4.1.   *Amorphous silicon p–i–n structures*

The basic a-Si solar cell is a *p–i–n* junction. Since diffusion lengths are so short in doped a-Si, the central undoped or *intrinsic* region is needed to extend the thickness over which photons may be effectively absorbed. The built-in bias is dropped across the width of the *i*-region, creating an electric field which drives charge separation (discussed in Sec. 5.5). In the *p–i–n* structure photocarriers are collected primarily by *drift* rather than by diffusion.

The thickness of the *i* region should be optimised for maximum current generation. Although more light is absorbed in a thicker region, charged defects reduce the electric field across the *i* region, and at some thickness the width of the *i* region will exceed the space charge width as shown in Fig. 8.10 below. The remaining, neutral part of the *i* layer is a 'dead layer' and does not contribute to the photocurrent.

Clearly the cell should be designed so that the depletion width is *greater than* the thickness of the *i* region at operating bias. In practice this limits

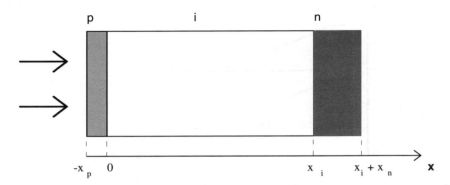

Fig. 8.9. Layer widths of p–i–n junction structure. Provided that the p and n layer doping levels are high enough, the depletion region is contained almost entirely within the i region.

the i-region thickness to around 0.5 μm. The following section explains how the current voltage characteristics of a p–i–n junction may be calculated.

8.4.2. *p–i–n solar cell device physics*

To a first approximation the *depletion approximation* can be used to calculate the current-voltage characteristics of a p–i–n junction solar cell. We will consider a structure with p, i and n layer thicknesses x_p, x_i and x_n and set $x = 0$ at the p–i interface. The p and n doping densities are N_a and N_d, respectively, and the i region has an unintentional or background doping level of N_i, due to charged impurities, here assumed n type. Because the doping levels in the p and n regions are so much higher than in the intrinsic region, the depletion widths within the p and n regions are very small and can be neglected. Then, provided that the background doping is low enough, the space charge region thickness is equal to the i-region thickness. Following the analysis of the p–n junction in Chapter 6, we have using Eq. 6.28,

$$J = -J_n(0) - J_p(x_i) - J_{\text{scr}} \qquad (8.1)$$

where J_{scr} is the current density from the i layer and each of the three contributions is worked out for some given bias V and generation rate G.

Because of the short diffusion lengths in doped amorphous silicon the current collection from the p and n layers is negligible. Moreover, because these layers are so thin, the excess recombination occurring there is also negligible compared to the recombination occurring in the intrinsic layer. So, to a first approximation, we need only include the photocurrent and dark current resulting from the intrinsic layer. Then, if dark current and photocurrent are independent, we have from Eq. 1.4

$$J(V) = J_{sc} - J_{dark}(V).$$

For the p–i–n structure, the dark current is given by the volume recombination rate integrated across the i region,

$$J_{dark}(V) = \frac{qn_i x_i}{\sqrt{\tau_n \tau_p}} \frac{2\sinh(qV/2kT)}{q(V_{bi} - V)/kT} \xi \tag{8.2}$$

where the Sah–Noyce–Shockley approximation, Eq. 6.43, has been used, and τ_n, τ_p refer to electron and hole lifetimes in the undoped region. (In fact, this is a far better approximation for p–i–n structures than for p–n's. The approximation supposes that the intrinsic energy level varies linearly across the depleted region, which is indeed the case for p–i–n structures with low background doping.) Because the depletion region is thick, i region recombination of the form Eq. 8.2 will always dominate and the ideal ($n = 1$) behaviour expected for diffusion currents in a p–n junction is seldom seen.

If collection of photogenerated carriers is perfect, the 'short circuit' photocurrent is given by the photogenerated current in the i region, from Eq. 6.44,

$$J_{sc} = \int q(1 - R)be^{-\alpha x_p}(1 - e^{-\alpha x_i})dE. \tag{8.3}$$

This is the case for very good carrier transport in the undoped region. If, however, carrier lifetimes are short or mobilities low, then the net photocurrent will be limited by the mobility and the electric field. The total photogenerated current, given by Eq. 8.3, may exceed the maximum drift current which the i-region can support at a given applied bias. In that case it can be shown that the net photocurrent is reduced to

$$J'_{sc} = J_{sc} \left(1 - \frac{x_i^2}{\mu\tau(V_{bi} - V)}\right) \tag{8.4}$$

where $\mu\tau$ is the average mobility-lifetime product for the two carrier types [Merten, 1998]. In deriving Eq. 8.4 it is assumed that electron and hole densities vary linearly with distance in the i region, and have independent,

constant lifetimes and that photogeneration is uniform. It is useful to define
a drift length for photogenerated carriers, L_i

$$L_i = \frac{\mu\tau(V_{bi} - V)}{x_i} .$$

Several semi-empirical models for performance of amorphous silicon solar
cells have been developed, which include such bias dependence of the photocurrent [*e.g.* Hegedus, 1997].

There are a number of ways in which this treatment may need to be
modified for a-Si p–i–n structures.

(i) If the *charged background doping* in the i region is too high, then at
biases above some threshold the i region will not be completely depleted.
This situation occurs when the i region thickness obeys

$$x_i \geq \sqrt{\frac{2\varepsilon_s(V_{bi} - V)}{qN_i}} \qquad (8.5)$$

where we have used Eq. 6.13 for the thickness of the depletion region at
a symmetric junction and assume that $N_a \gg N_i$. As V is increased the
built-in bias is split between the p–i and i–n interfaces, so that in the case
of n type background doping in the i region, only

$$V_{pi} = \frac{kT}{q} \ln \left(\frac{N_a N_i}{n_i^2} \right)$$

is dropped across the p–i depletion region and the depletion thickness is
reduced to

$$w_i = \sqrt{\frac{2\varepsilon_s(V_{bi} - V)}{qN_i}} . \qquad (8.6)$$

Then in Eqs. 8.2 and 8.4 V_{bi} should be replaced by V_{pi} and x_i should be
replaced by w_i from Eq. 8.6. The situation is illustrated in Fig. 8.10.

In this situation the photocurrent will be reduced and become bias dependent, decreasing with increasing V. For high N_i levels this effect appears
as a slope in the plateau of the J–V curve, and could be confused with shunt
resistance.

(ii) At *high injection* levels, for instance at high illumination, the free
carrier densities in the i region become significant and the depletion approximation is not valid for calculating the band profile. The band profile
should then be calculated self consistently, including the contributions of n

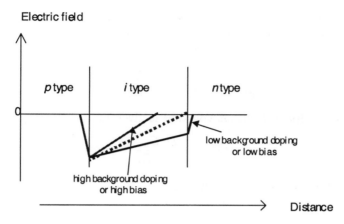

Fig. 8.10. Electric field dropping across an *i* layer with charged background impurities. The dashed line represents the electric field distribution at the point where the *i* region is just depleted.

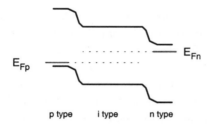

Fig. 8.11. Band profile of *p–i–n* junction under high injection conditions. The electric field in the *i* region is neutralised by the space charge of free carriers.

and *p*. Free carriers arrange themselves to minimise the electrostatic potential energy with the result that the electric field, rather than being constant across the *i* region, is higher at the *p–i* and *i–n* interfaces, and smaller in the middle. In very high injection conditions the field vanishes throughout most of the *i* region and $n \approx p$.

Then *n* and *p* obey equations of the form

$$\frac{d^2p}{dx^2} - \frac{(p - p_0)}{D_p \tau_p} + \frac{g(E, x)}{D_p} = 0 \tag{8.7}$$

$$\frac{d^2n}{dx^2} - \frac{(n - n_0)}{D_n \tau_n} + \frac{g(E, x)}{D_n} = 0 \tag{8.8}$$

where D_n, D_p, are the diffusion coefficients of electrons and holes in the intrinsic region. Since $n = p$ and $n_0 = p_0$, these equations can be added to give an *ambipolar* diffusion equation

$$\frac{d^2n}{dx^2} - \frac{(n - n_0)}{L_0^2} + \frac{g(E, x)}{D_a} = 0 \qquad (8.9)$$

where the ambipolar diffusion constant is given by

$$D_a = \frac{D_n D_p}{D_n + D_p} \qquad (8.10)$$

and the ambipolar diffusion length by

$$L_a = \sqrt{D_a \tau_a} \qquad (8.11)$$

where

$$\tau_a = \frac{D_n + D_p}{D_n/\tau_p + D_p/\tau_n} . \qquad (8.12)$$

Evidently the current is now driven by diffusion rather than drift, and charge separation is driven by the small space charge regions for holes at the p–i interface and for electrons at the i–n interface.

In these conditions the recombination rate becomes linear with n (the SRH recombination rate, Eq. 4.79, becomes approximately proportional to n when $n = p$) and is constant across the i region. Then the ideality factor of the dark J–V characteristic should become equal to one. In the case where electron and hole lifetimes are equal, and n, $p \gg n_i$, Eq. 8.9 can be solved exactly to give solutions analogous to those for the currents from the neutral regions of a p–n junction derived in Chapter 6.

(iii) The electron and hole lifetimes in Eq. 8.2 are, in general, not constants for an amorphous material. At low light intensity charge carriers are more likely to be trapped in tail states, which extends their lifetime. The *dark* current term in Eq. 1.3 is therefore *intensity* dependent. The effect for amorphous silicon p–i–n cells is that performance is better at low light intensity. Carrier density dependence of lifetime is evident through decreasing collection efficiencies and fill factor with increasing light intensity.

(iv) If the electron and hole mobilities are low, collection in a wide, depleted i layer may be limited by the electric field due to the mobile carriers. In these conditions the current is said to be 'space charge limited'. If mobility is intensity dependent, collection in this limit should improve with increasing light intensity.

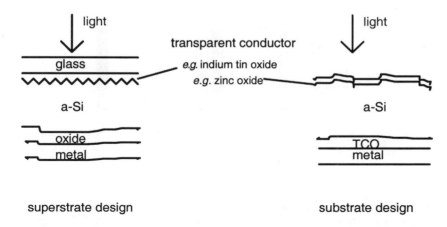

Fig. 8.12. Substrate and superstrate a-Si cell designs.

8.4.3. *Fabrication of a-Si solar cells*

Amorphous silicon solar cells are normally deposited on glass substrates, which are coated with a transparent conducting oxide (TCO) such as tin oxide or indium tin oxide. TCO coated plastics are also being developed. Cells are usually fabricated in a 'superstrate' design, where layers of conducting oxide, p-type, undoped and n-type a-Si are deposited in sequence. The a-Si is usually deposited by plasma decomposition of silane or 'glow discharge', but a number of other deposition methods such as sputtering and 'hot wire' are being investigated. For the rear contact, zinc oxide is deposited on to the n layer followed by a metal, usually aluminium. Light trapping structures may be built-in by texturing the front TCO layer, and metallising or texturing the back surface in order to enhance light absorption.

An alternative is the 'substrate' design where layers are deposited on a metal substrate, such as steel, which forms the back contact. Here the p layer can be very thin, since there is no glass and the front surface does not need to be flat, and so higher efficiencies are possible. However, the substrate design is not so easy to process.

8.4.4. *Strategies to improve a-Si cell performance*

- Light induced degradation

The Staebler Wronski effect is the most important barrier to widespread use of a-Si solar cells. Light-induced degradation is stronger

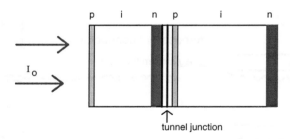

Fig. 8.13.   Multilayer a-Si cell.

in materials with a higher hydrogen content, because of the greater density of Si–H bonds, yet a high hydrogen content is needed for suitable doping and transport properties. If a-Si could be produced with a lower original dangling bond density then less hydrogen would be needed for passivation and so the Staebler Wronski effect could be reduced. Alternative deposition techniques are therefore being studied. One possibility is the 'hot wire' technique which appears to produce good a-Si when saturated with only 1% hydrogen.

- Improvement of V_{oc}

   The open circuit voltage in a-Si solar cells is substantially less than the optical band gap (0.89 V compared to 1.7 eV) on account of the high activation energies in amorphous material, and resulting low built-in bias. V_{oc} can be increased by the use of either (i) a wider band gap emitter such as a-SiC:H or (ii) a polycrystalline Si emitter, in which degenerate doping is possible.

- Improvement of J_{sc}

   There are two problems:

   (i) Response to blue light is poor in homogenous a-Si:H cells on account of poor collection in the p layer. This can be resolved by replacing the a-Si p layer with a wider band gap a-SiC:H window, like the heteroface designs preferred for GaAs, or by a graded a-SiC:H-a-Si:H layer.

   (ii) Response to long wavelengths may be poor because of the limit to the i-region thickness which arises from charged background doping in the i-region. This can be resolved either with light trapping techniques to increase the optical path length within the cell, or with

Table 8.1.   Evolution of performance of a-Si solar cells.

Date	Design	Efficiency
1977	Schottky diode	6%
1980	a-Si:H $p-i-n$	6%
1982	a-SiC:H/a-Si:H $p-i-n$ heterojunction	8%
1982	textured substrates	10%
1987	grading of $p-i$ interface	12%
1990s	multigap designs	13%

the use of multilayer a-Si cells. In the multilayer design, two or more $p-i-n$ cells with relatively thin i-regions are connected in series. The extra cell (or cells) increases overall light absorption while the thin i regions avoid the collection losses due to background doping in the i region. Moreover, the higher electric field which applies in thinner i layers appears to reduce the Staebler Wronski effect.

Since the cells are connected in series, it is necessary to match the currents from the front and back cell. Since less light reaches the back cell, the back cell must be thicker to produce the same photocurrent. The optimum ratio of thicknesses will depend on illumination conditions. As with two-terminal tandem designs (discussed in Chapter 10), the stack will operate at its optimum only under certain illumination conditions.

- Improvement of limiting efficiency

  Multi-gap cell designs are possible using a-SiC:H as the material for a wider gap cell and a-SiGe:H for a narrower gap cell. Two-terminal cascade designs where the different $p-i-n$ cells are connected in series using tunnel junctions have been studied. To date the best efficiency from a three-cell stack slightly improves on the best single junction a-Si cell, but substantial improvements are possible. The limiting efficiency for three cell devices is calculated at 33%. The main problems have been the poorer quality of the alloy relative to a-Si and incorporating large area tunnel junctions. The evolution of a-Si solar cell performance is charted in Table 8.1.

8.5.   Defects in Polycrystalline Thin Film Materials

The following sections are concerned with polycrystalline thin film photo-voltaic materials. We first consider the features of a grain boundary and the consequences for photovoltaic devices.

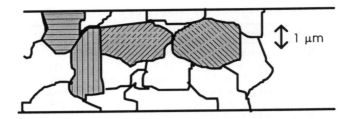

Fig. 8.14.   Grain boundaries in microcrystalline material.

8.5.1.   *Grain boundaries*

A polycrystalline material is composed of microcrystallites or 'grains' of the semiconductor arranged at random orientations to each other (Fig. 8.14). The material is crystalline over the width of a grain, which is typically the order of one μm. Since the grains are large in quantum-mechanical terms, the band structure, and therefore the absorption coefficient, is virtually identical to that of the single crystal material. However the transport and recombination properties are strongly affected by the presence of the interfaces or *grain boundaries*.

The different orientations of neighbouring crystal grains give rise to dislocations, misplaced atoms ('interstitials'), vacancies, distorted bond angles and bond distances at the interfaces. In compound semiconductors we may find atoms occupying the wrong lattice sites, and vacancies of particular atoms which effectively dope the material. There may be a few atomic layers of imperfect crystal at the grain boundaries, while the material accommodates the change in crystal orientation (Fig. 8.15). Polycrystalline materials are also likely to contain *extrinsic* impurities which contaminate the materials during growth. These extrinsic impurity atoms are likely to concentrate at the grain boundaries.

The various types of defect introduce extra electronic states. These extra states are spatially localised and because they do not need to obey the symmetry of the crystal and they may have energies in the band gap (see Chapter 5), these intra band gap states tend to trap carriers and we will refer to them as 'intra-band-gap states' or simply 'trap states'. How they behave depends upon their energetic position relative to the bands of the bulk crystal: shallow defects close to the conduction band tend to act as electron traps (acceptors) while defect levels close to the valence band act as hole traps (donors). 'Deep' defect levels at energies near the centre of the

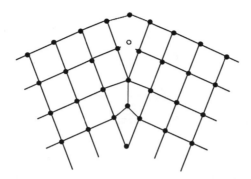

Fig. 8.15. Schematic of grain boundary showing distorted lattice between grains of different orientation.

band gap may be capable of trapping either type of carrier, and because the times for re-release from deep traps by thermal activation are very long they tend to act as recombination centres. (See Sec. 4.5.)

Because these intra band gap states are able to trap charge, they influence the potential distribution close to the grain boundary. In Chapter 5 we saw how defect states at a surface in doped material tend to trap majority carriers and establish an electrostatic field opposing majority carrier flow. The situation with a grain boundary is similar, rather like two surfaces back to back. We consider the case of a grain boundary in n type material in Fig. 8.16. The defect levels within the band gap are usually distributed so that the local neutrality level (defined in Chapter 5 as the level up to which the states are filled when the interface is neutral) lies closer to the centre of the band gap than the Fermi level of the doped semiconductor. (This will always be the case unless the interface states are concentrated very close in energy to the conduction band edge.) Then, the states will be acceptor-like and trap electrons. This gives rise to a plane of fixed negative charge at the interface, and a layer of positive space charge on either side where the n type material had been depleted. The electrostatic force sets up a potential barrier which opposes further majority carrier migration. *Minority* carriers, however, see a potential well at the grain boundary and are pulled towards it, where the probability of recombining with a trapped majority carrier is high.

Figure 8.16 shows how a depletion region is established around a grain boundary in n type material for which $\phi_0 < E_F$. The grain boundary

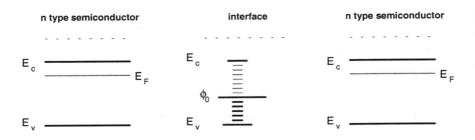

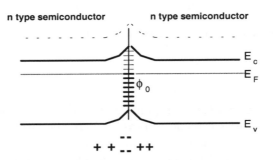

Fig. 8.16.   Illustration of how band bending results when a layer of interface states is sandwiched between two n type semiconductor layers. Notice how depletion results when the neutrality level of the interface states is lower than the Fermi level of the nuetral n type layer.

accumulates a (negative) sheet charge density of

$$Q_{gb} = -q \int_{\phi_0}^{E_c - V_n - E_B} g_{gb}(E) dE \qquad (8.13)$$

where $g_{gb}(E)$ is the density of interface states per unit area, V_n is the donor ionisation energy, and E_B is the band bending due to the trapped charge, found from Poisson's equation. The depleted layers of semiconductor on either side should each contain a compensating (positive) charge density of $+\frac{1}{2}Q_{gb}$.

For p type material the situation is analogous. Occupied defect levels below ϕ_0 trap majority holes until a barrier opposing hole flow is established, and a negatively charged depleted region appears around the positively charged grain boundary.

8.5.2. *Effects of grain boundaries on transport*

The effect of grain boundaries on charge transport depends on whether they lie normal to or parallel to the direction of current flow. In the first case, when current is flowing *across* a grain boundary, the potential barriers slow down the transport of majority carriers, limiting the majority carrier mobility, while the potential wells drive minority carriers towards recombination centres at the grain boundary, reducing the minority carrier diffusion length and lifetime. The size of these effects depends upon the doping, the density of interface states, and the photogenerated carrier density. A simple model for a grain boundary in n type materials [Seto, 1975; Card, 1977; Landsberg, 1984] is discussed below.

The main results are that:

- Increasing the trap density increases the space charge stored at the grain boundary, which increases the barrier height, reduces conductivity and increases recombination.
- Increasing the doping first increases the barrier height, but at higher doping levels the traps become saturated, the space charge region begins to contract and the barrier is reduced.
- Increasing the density of free carriers by illumination reduces the net charge stored at the grain boundary and hence the barrier height. At high illumination levels the grain boundaries have the minimum effect: conductivity reaches its maximum level, and recombination with trap states saturates.

Grain boundaries which lie parallel to the direction of current flow principally affect minority carriers. Majority carriers travelling parallel to the

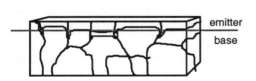
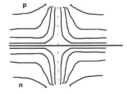

Fig. 8.17.   (a) In a microcrystalline material, the intended width of the emitter region may be comparable with the grain size. In such cases, the p–n junction passes through individual grains and the actual position of the junction (grey line) may vary from grain to grain; (b) At a point where the p–n junction (full line) passes through a grain boundary (dashed line) the local potential is affected, as shown by the lines of constant electrostatic potential (curved lines).

grain boundary are not affected — they see no barrier — but minority carriers are still likely to be trapped in the potential well and recombine. When a grain boundary actually crosses the p–n junction, it reduces the efficiency of charge separation by competing with the p–n junction for minority carriers. A more serious problem arises when dopant impurity atoms from the emitter diffuse along the grain boundaries, through the nominal p–n junction, into the base. This creates a shunt path through the p–n junction which reduces the rectification of the junction.

8.5.3. *Depletion approximation model for grain boundary*

Here we are going to use the depletion approximation to show how the resistivity of a granular material depends upon doping, density of trap states and light intensity. The grain boundary is modelled as a plane at $x = 0$ with sheet density N_s of defects sandwiched between two layers of a homogenous material degenerately doped n type with dopant density N_d. Neighbouring grain boundaries are placed at $x = \pm d$. The defect energy levels are located in a narrow band around E_t such that E_t is less than the donor level $E_c - V_n$. In isolation, the neutrality level of the grain boundary is below the Fermi level of the bulk semiconductor as shown in Fig. 8.16. When the three layers are brought together, electrons move from the bulk layers into trap states at the grain boundary until equilibrium is reached. Then the plane of the grain boundary will be negatively charged and the surrounding bulk layers positively charged, resulting in the potential profile shown in Fig. 8.18. The conduction band edge at the grain boundary is raised by an amount E_B relative to its value at the centre of the grain.

In the depletion approximation we neglect free carriers and suppose that the bulk layers are completely depleted for a thickness L on either side. Then solving Poisson's equation with the boundary condition that the electric field vanishes at $x = \pm L$ field we find that the conduction band edge varies like

$$E_c(x) = E_c(L) + \frac{q^2 N_d}{2\varepsilon_s}(L - |x|)^2 \,. \tag{8.14}$$

Now we need to distinguish two cases:

(i) the grain is completely depleted and the trap states are partly filled. ($dN_d < N_s$) This occurs at low doping.

(ii) the grain is partly depleted and the trap states are completely filled ($dN_d > N_s$). This occurs at high doping or at high intensity levels.

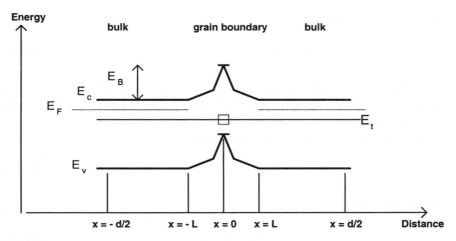

Fig. 8.18. Grain boundary in the depletion approximation, showing trap energy E_t, barrier height E_B and Fermi level E_F.

In the first case $L = \frac{d}{2}$ and the barrier height is given by

$$E_B = \frac{q^2 N_d d^2}{8\varepsilon_s}. \tag{8.15}$$

In this case, because the grain is completely depleted, the Fermi level has been *unpinned* from the donor level of the bulk material. At the middle of the grain the Fermi level has been shifted down by an amount Δ and the electron density reduced to $N_d e^{-\Delta/kT}$ relative to the undepleted material. To find the new value of E_F we need to consider the filling of the trap levels. The sheet density of occupied trap states is given by

$$dN_d = \sum_{\text{traps,t}} f(E_t, E_F, T) g_{gb}(E_t) \tag{8.16}$$

where $f(E_t, E_F, T)$ is the Fermi–Dirac distribution function (Eq. 3.25). With a delta function for the distribution of trap states per unit area

$$g_{gb}(E) = N_s \delta(E - E_t) \tag{8.17}$$

and assuming a spin degeneracy of one we find

$$dN_d = \frac{N_s}{e^{(E_t - E_F)/k_B T} + 1}. \tag{8.18}$$

Rearranging, we find for E_F

$$E_F = E_t - kT \ln \left(\frac{N_s}{dN_d} - 1 \right) \tag{8.19}$$

where E_F and E_t are measured relative to some reference, such as the intrinsic level at $x = 0$. Thus the Fermi level is unpinned from the donor level, and fixed instead to the trap energy level.

In the second case, where the grain is partly depleted,

$$L = \frac{N_s}{2N_d}, \tag{8.20}$$

the barrier height is

$$E_B = \frac{q^2 N_s^2}{8\varepsilon_s N_d} \tag{8.21}$$

and the conduction band minimum is at the conduction band edge energy of the bulk material in equilibrium. In the mid-grain, undepleted regions the Fermi level is pinned at the donor level and n has its bulk value N_d. The charge distributions, electric field and band profiles are illustrated for each case in Fig. 8.19 below.

8.5.4. *Majority carrier transport*

In studying the transport of majority carriers we are interested in the effective *conductivity* of the medium, which is the current crossing the grain boundary per unit applied field, and the effective *mobility*, which is the conductivity per unit charge. For this we need to calculate the mean current density crossing the grain boundary, $\langle J \rangle$.

We suppose that a constant electric field F is applied to the granular medium of the last section. Then the electrostatic potential is on average higher by Fd on the right than the left of the grain boundary in Fig. 8.20, and a net current of approximately

$$\langle J \rangle = q \langle n \rangle v_x e^{-E_B/k_B T} (e^{qFd/k_B T} - 1) \tag{8.22}$$

will flow to the right. Here $\langle n \rangle$ is the mean electron density and v_x is the mean electron velocity in the x direction, given by

$$v_x = \left(\frac{k_B T}{2\pi m_c^*} \right)^{1/2} \tag{8.23}$$

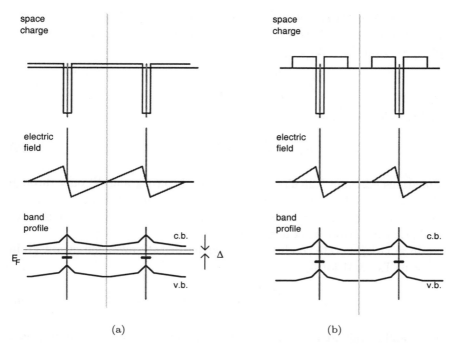

Fig. 8.19. Space charge, electric field and band profile at a grain boundary in *n*-type material for the case where the grains are (a) totally and (b) partly depleted. In case (a) the Fermi level, represented by the full horizontal line in the band profile diagram, is shifted downwards relative to the donor level (dashed line) by an amount Δ. In case (b) the Fermi level is pinned at the donor level.

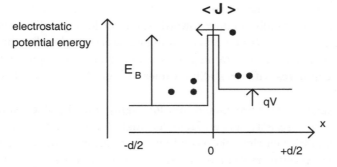

Fig. 8.20. Model of thermionic current across a grain boundary under electric field V/d.

assuming Boltzmann statistics and isotropic motion, where m_c^* is the electron effective mass. Since Fd is small compared to $k_B T/q$, the conductivity obtained from Eq. 8.22 is given by

$$\sigma = \frac{\langle J \rangle}{F} = \frac{q^2 \langle n \rangle v_x d}{kT} e^{-E_B/k_B T} \tag{8.24}$$

and the electron mobility by

$$\mu_n = \frac{\sigma}{q \langle n \rangle} = \frac{q v_x d}{kT} e^{-E_B/k_B T} \, . \tag{8.25}$$

Equations 8.24 and 8.25 show that σ depends on N_d and N_s only through E_B and $\langle n \rangle$, and μ_n only through E_B. It is immediately clear that μ_n will be lowest where E_B is greatest. To consider the behaviour of σ we need to evaluate the mean carrier density $\langle n \rangle$.

In non-degenerate conditions the carrier density averaged over a grain is, using Eq. 3.31 for n,

$$\langle n \rangle = \frac{2}{d} \int_0^{d/2} N_c e^{-(E_c(x)-E_F)/k_B T} dx \, . \tag{8.26}$$

This can be evaluated exactly using Eq. 8.14 for $E_c(x)$ with the appropriate values of L in each case. The result has the limiting forms

$$\langle n \rangle \approx n_i e^{(E_B+E_F)/kT} \tag{8.27}$$

for $dN_d \ll N_s$ and

$$\langle n \rangle \approx N_d \tag{8.28}$$

for $dN_d \gg N_s$. This has the consequence that σ is small and insensitive to doping over the regime where $dN_d < N_s$, but then rises rapidly with N_d once $dN_d > N_s$, finally tending towards the limit for crystalline material as $E_B \to 0$. The *resistivity* of a polycrystalline material is therefore high at low doping levels, falling very rapidly at some intermediate doping levels until finally it compares with crystalline values at high N_d. This is illustrated in Fig. 8.21.

This suggests that we should aim for highly doped polycrystalline material to reduce resistive losses. However, high doping tends to increase the losses of minority carriers via recombination at grain boundaries, as will be seen below. As with crystalline materials, the best material is a compromise between high doping and high conductivity.

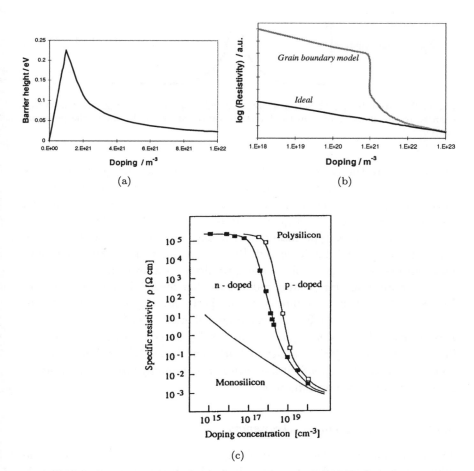

Fig. 8.21. Grain boundary model of resistivity in polycrystalline material. (a) Barrier height as a function of doping density, from Eqs. 8.15 and 8.21. The maximum occurs when $dN_d = N_s$. (b) Resistivity for grain boundary calculated from Eqs. 8.24 and 8.27, compared with the case of an ideal crystal with $E_B = 0$. Parameters used were $N_s = 10^{15}$ m^{-2}, $d = 1$ μm, $\varepsilon_s = 10\varepsilon_0$. (c) Data for the resistivity of polycrystalline and crystalline silicon [Moller, 1983].

8.5.5. *Effect of illumination*

By analogy with the case of bulk recombination via a single trap level (Eq. 4.78), the fraction of interface traps which are occupied is given by

$$f_t = \frac{S_n n - S_p p_t}{S_n(n + n_t) + S_p(p + p_t)} \tag{8.29}$$

Fig. 8.22. Grain boundary height at different light intensities. At high light intensity, changes in the charge stored in the interface states reduce the height of the potential barrier.

where n, p are the densities of electrons and holes at the grain boundary and S_n, S_p are the interface recombination velocities. Illumination increases both n and p, but the fractional increase in the minority carrier population — holes in this case — is greater. It is clear from Eq. 8.29 that, for a grain boundary in n type material, if n and p are both increased by the same amount, f_t will be decreased. Physically, this is due a greater rate of hole capture by the grain boundaries, resulting in a lower net charge. The reduced grain boundary charge reduces the barrier height E_B, as shown in Fig. 8.22. At first, if the minority carrier Fermi level is still above the trap level (*i.e.*, $p < p_t$) the effect is negligible and E_B is unaffected. Then, as p surpasses p_t, the barrier begins to fall. As E_B is reduced under increasing illumination the majority carrier mobility, and hence the conductivity, are increased. Intensity dependent mobilities are commonly observed in practice in polycrystalline semiconductors.

8.5.6. *Minority carrier transport*

The most important factor controlling minority carrier transport is the minority carrier recombination rate. Recombination at a grain boundary is analogous to surface recombination and can be treated with the approach of Sec. 4.5. The recombination current density, J_{gb}, can be expressed as q times the area recombination rate at the grain boundary, using Eq. 4.82,

$$J_{gb} = qU_{gb}\delta x = q\frac{np - n_i^2}{\frac{1}{S_n}(p + p_t) + \frac{1}{S_p}(n + n_t)} \qquad (8.30)$$

where n and p are the electron and hole densities at the grain boundary. Now because of the depletion region, the electron density at the grain boundary will be *lower* than its bulk value by approximately e^{-E_B/k_BT} and the hole density *higher* than its bulk value by e^{E_B/k_BT}, so the net recombination rate is higher than it would be in the absence of the depletion region, and is not simply proportional to the minority carrier density as expected for bulk recombination in a doped material (Eq. 4.80). However, it is convenient to define an *effective* interface recombination velocity, S_{gb},

$$S_{gb} = -\frac{J_p(L)}{q(p(L) - p_0)} \tag{8.31}$$

where $J_p(L)$ and $p(L)$ are evaluated at the edge of the depletion layer and p_0 is the equilibrium hole density. $J_p(L)$ is equal to the interface recombination current density, whence

$$J_{gb} = J_p(L) = -qS_{gb}(p(L) - p_0). \tag{8.32}$$

At equilibrium, net grain boundary recombination is zero. Under increasing illumination, the net charge at the grain boundary decreases so that the barrier height decreases, and the *enhancement* of the recombination rate due to the depletion region decreases. The effect is that, at intensities such that $p > p_t$, the ratio in Eq. 8.31 decreases, and the effective recombination velocity decreases.

To evaluate the overall effect on the minority carrier transport characteristics we need to add the grain boundary recombination to the other principal recombination mechanisms. We define a minority carrier lifetime for grain boundary recombination from the grain boundary recombination rate *per grain*, (*i.e.*, a volume recombination rate)

$$\frac{1}{\tau_{gb}} = \frac{J_{gb}}{qL(p(L) - p_0)} = \frac{S_{gb}}{L}. \tag{8.33}$$

Then an effective minority carrier lifetime, τ_{eff}, can be defined by adding grain boundary recombination to the other mechanisms,

$$\frac{1}{\tau_{eff}} = \frac{1}{\tau_{SRH}} + \frac{1}{\tau_{gb}} + \frac{1}{\tau_{Auger}} + \frac{1}{\tau_{rad}}. \tag{8.34}$$

An effective diffusion length L_{eff} may be derived from τ_{eff} in the usual way, as for Eqs. 4.95 and 4.96

$$L_{eff} = \sqrt{D_{eff}\tau_{eff}}. \tag{8.35}$$

In conditions where grain boundary recombination dominates — that is, low doping, high grain boundary defect density and low illumination — the overall effective lifetime and diffusion length will be bias and intensity dependent, through the bias dependence of S_{eff}. At higher intensity levels the other mechanisms take over, and L_{eff} increases towards a saturation

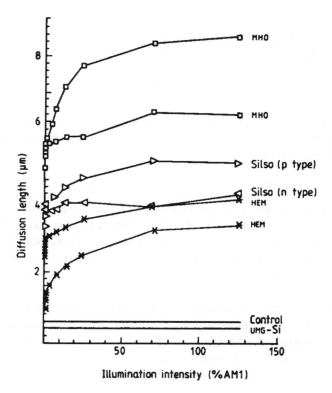

Fig. 8.23. Increase in effective diffusion length L_{eff} with carrier density in p-Si. [de Pauw, 1984.]

value. In polycrystalline silicon, L_{eff} has been observed to increase with generation rate up to a saturation level (Fig. 8.23). In that material grain boundary effects dominate for doping levels $N_a < 10^{17}$ cm^{-3}.

8.5.7. *Effects of grain boundary recombination on solar cell performance*

Grain boundary effects influence the current voltage characteristic of poly-crystalline solar cells in several important ways.

- The effect of grain boundaries in reducing majority carrier mobility may increase series resistance.
- The enhancement of minority carrier recombination at grain boundaries reduces lifetimes, as discussed above, and increases dark current. Notice

that since grain boundary recombination occurs in space charge regions where n and p are comparable, we tend to see the dark current characteristic typical of generation-recombination currents (with ideality factor of 2) if the grain boundary defect density is high enough.

- Because minority carrier lifetimes and diffusion lengths are carrier density dependent, the reverse current which opposes the short circuit photocurrent may be intensity dependent whilst the *photocurrent* may be bias dependent, and not well approximated by J_{sc}. These mean that simple diode equation descriptions such as Eq. 1.4 are not appropriate, unless grain boundary effects are well controlled.

A further influence on J–V characteristics arises from recombination at the interface in heterojunction cells, discussed below.

8.6. CuInSe$_2$ Thin Film Solar Cells

8.6.1. *Materials properties*

Copper indium diselenide (CuInSe$_2$, or CIS) is a direct gap semiconductor with a band gap of around 1 eV and an optical absorption which is amongst the highest known for any semiconductor. Its strong absorption, as well as its availability in both p and n types, make it attractive for thin film photovoltaics and it has been developed for this purpose since the early 1970s. In its polycrystalline form, it consists of grains of size approximately 1 μm, which tend to grow in a columnar structure so that grain boundaries are more likely to lie normal to the p–n junction. CIS p–n homojunction solar cells can be made, but have an efficiency of only 3–4%, and better devices can be made with a *heterojunction* structure, using an n-type CdS emitter on a lightly doped p-type CIS base. The highly doped CdS layer has a band gap of around 2.5 eV and serves (i) as a window layer to reduce the collection losses due to surface recombination of carriers photogenerated by short wavelength light, and (ii) to transport electrons from junction to front surface with minimum series resistance. Only a few μm of p-CIS base are needed for good absorption. In most photovoltaic structures the alloy CuInGaSe$_2$ (CIGS) is used in place of CIS. Addition of Ga improves the photovoltaic characteristics, by raising the band gap, as well as the electronic properties of the rear contact.

The compound forms the chalcopyrite crystal structure. The conductivity of p type CIGS is due to native defects, mainly indium vacancies and copper atoms on indium sites, and can be controlled by varying the Cu/In ratio during growth. These native defects tend to congregate at the grain boundaries and give rise to a distribution of trap energies above the valence band edge such that the hole activation energy is around 0.3 eV. (Note this is smaller than the value for a-Si, indicating that charge compensation is less of a problem.) In addition, positively charged selenium vacancies concentrate at the grain boundaries and must be passivated, for instance by annealing in oxygen, to reduce grain boundary activity. Electron diffusion lengths are similar to the grain size, giving rise to internal quantum efficiencies which are as high as 90% at the maximum.

8.6.2. *Heterojunctions in thin film solar cell design*

In a heterojunction the emitter is made from a different, wider band gap material than the base. Such designs are used in polycrystalline thin film solar cells for two reasons. First, the short diffusion lengths mean that high energy photons may not be collected efficiently. The wider band gap emitter therefore acts like a window (see Sec. 7.6). The second reason is that CdTe and CuInSe$_2$ thin film materials cannot easily be doped n-type, so alternative materials such as CdS must be used.

The material interface introduces several problems:

(i) Differences between the crystal structure or lattice constants of the two materials generally introduce intra band gap defect states at the junction. These states encourage Shockley Read Hall recombination, increasing dark currents and limiting the open circuit voltage. They may also assist *tunnelling* of carriers across the junction, again increasing recombination. Very large densities of interface states can lead to Fermi level pinning at the interface, with the effect that carrier densities are not easily controlled.

(ii) Differences between the chemical composition of the two materials may lead to the diffusion of species across the junction, or the formation of new chemical compounds in the junction region. For example, the unwanted alloys CuSe$_2$ and CuS$_2$ may form at the CuInSe$_2$ — CdS heterojunction.

(iii) Differences between the electron affinity and band gap of the two materials may lead to the situation where there is a narrow barrier or spike in the conduction or valence band, or both, at the interface. Then photogenerated carriers crossing the junction need to tunnel through the spike before they can be collected, and this reduces the

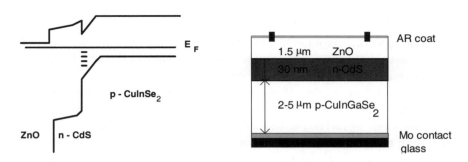

Fig. 8.24. Band profile and layer structure of typical CIGS cell.

efficiency of collection. At the Cu(In,Ga)Se$_2$ — CdS junction, a step of around 0.5 eV in the conduction band edge creates a triangular barrier (Fig. 8.24(a)), opposing collection of photogenerated electrons. This step acts like a Schottky barrier, but its effect on photocurrent collection is negligible if it is less than 0.4 eV. It introduces a contribution to the recombination current from tunnelling of electrons to interface defects in the CdS, and this is manifest as an ideality factor greater than 2 in the (low temperature) dark current characteristic [Rau, 2001]. The problem of a barrier at the heterojunction is avoided for CdS — CdTe n–p structures.

8.6.3. *CuInGaSe$_2$ solar cell design*

CIGS is deposited by two main methods: vapour co-deposition of copper, indium, gallium and selenium; and the selenisation of Cu/In films. After deposition of the CIGS layer the film is annealed in oxygen, to compensate charged grain boundary defects and reduce the density of recombination centres. Then the n type CdS layer is deposited typically by chemical bath deposition, followed by a thin layer of conducting oxide which may also serve as part anti-reflection coat. Zinc oxide is preferred as its high conductivity reduces the thickness of CdS needed to a few tens of nm. CIGS cells are always prepared as substrate designs on a molybdenum coated glass substrate, as Mo is needed to make an Ohmic contact. Ga improves adhesion to the Mo contact, but is costly.

The efficiency of CIGS cells improved substantially over the last 10 years and presently stands at 18.4% for a device on glass substrate [Contreras, 1999].

The major design problems with CIGS cells are:

(i) Poor spectral response at long wavelengths, which is due to the poor collection of carriers photogenerated at the back of the cell. Improved photocurrent requires a longer electron diffusion length, and since L_n at about 2 μm is already similar to the grain size this means reducing grain boundary recombination. This may be done by passivating the interfaces, or increasing the crystallite size by annealing.

(ii) High recombination at the heterojunction, and grain boundaries. The dark current is about ten times that of crystalline silicon (which has a similar band gap). The high defect density at the heterojunction arises from the mismatch of the wurtzite crystal structure of CdS with the chalcopyrite structure of CIGS, and encourages recombination by tunnelling via trap states.

(iii) Although optical losses in the CdS can be minimised by keeping that layer as thin as possible, poor collection in the blue and green is a problem with mass-produced cells, because of poorer surface processing.

8.7. CdTe Thin Film Solar Cells

8.7.1. *Materials properties*

Cadmium telluride (CdTe) is the semiconductor from the II–VI group of materials which is in principle best suited for photovoltaics. It has a direct band gap of 1.44 eV, close to the optimum for photoconversion, and a very high optical absorption. It is one of only two II–VI compounds (the other one is ZnTe, with a band gap of 2.26 eV) which can be doped p type as well as n type. Historically, CdTe was sought and developed as a p type replacement for the photovoltaic material, Cu_2S, which was used in early heterojunction designs with n type CdS as the wider band gap emitter, but whose performance was inhibited by diffusion of Cu along the grain boundaries.

The crystal adopts the wurtzite crystal structure, like GaAs, but has poorer transport properties. It suffers from a high density of native defects such as excess Te atoms at the grain boundaries which give rise to defect states deep in the band gap. p type CdTe must be doped extrinsically but the high intra band defect density leads to poor doping efficiency, much as in amorphous silicon. However, treatments have been developed to saturate the traps and efficient (> 15%) solar cells have now been produced from the polycrystalline material. Activation by heating in $CdCl_2$ is essential to

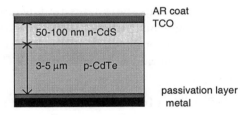

Fig. 8.25. Layer structure of CdTe cell.

improve material quality. Grain size can be increased and material properties improved by annealing.

CdTe can be single crystal or polycrystalline, although polycrystalline is obviously the material of interest for thin film cells and in fact it has achieved the best performance. A wide variety of techniques have been used for the deposition of CdTe in solar cells. The best results to date have been achieved with close space vapour transport, but other candidates include gas phase evaporation and spray pyrolysis.

8.7.2. *CdTe solar cell design*

The preferred design is a n-CdS p-CdTe heterojunction cell. As in CuInSe$_2$ based cells the CdS emitter acts as a window to improve collection at short wavelengths. In this case, no barrier results in the conduction or valence band, but lattice mismatch between the two materials leads to interface states. About 3–5 μm of CdTe are needed for sufficient optical depth. This is generally weakly doped but a second doping treatment, *e.g.* with copper or lithium is needed to improve the conductivity of the CdTe layer at the rear contact. The materials may be deposited in substrate fashion on a metal substrate, or in superstrate fashion on TCO coated glass.

Problems facing CdTe cell design are as follows:

- The CdS emitter still absorbs significantly in the green and very thin (50–80 nm) layers are needed for good response to green light. One objective in CdTe cell development is to reduce the thickness of the CdS layer in mass production.
- The high density of trap states at the grain boundaries give rise to a high dark current (about 20 times that of the otherwise similar material GaAs) and low V_{oc} (0.2 V less than GaAs). The low V_{oc} is partly attributed to Fermi level pinning at the trap states.

- Fermi level pinning by the trap states leads to difficulties doping CdTe heavily p type and making Ohmic contact to the substrate.
- Defects at the CdS-CdTe heterojunction, due to lattice mismatch or to the formation of other chemical compounds (such as $CdTeO_3$) enhance junction recombination.

The most efficient CdTe cell is a 16.4% efficient, 1 cm^2 CdS/CdTe device on glass, produced by the US National Renewable Energy Laboratories. The challenge is now to improve CdTe solar cell production technology so as to increase the efficiency of mass produced cells towards the values for lab cells.

8.8.   Thin Film Silicon Solar Cells

8.8.1.   *Materials properties*

Thin film microcrystalline silicon is characterised by grain sizes of around 1 μm. It has the optical properties of crystalline silicon, while its electronic properties are dominated by the grain boundaries. Defect states at grain boundaries include dangling bonds typical of amorphous silicon as well as extrinsic impurities introduced during growth. The defect states appear to be distributed through the band gap so that the grain boundaries are active in both n and p type material. Grain boundaries dominate the transport properties of microcrystalline Si for doping levels $N_a < 10^{17}$ cm^{-3}, giving rise to intensity dependent minority carrier diffusion lengths and mobilities.

Microcrystalline silicon (μ-Si) can be prepared by the same techniques as multi-crystalline, normally by casting of molten silicon into aggregates or sheets. A range of other techniques have been investigated for thin film polycrystalline material, including liquid phase epitaxy, chemical vapour deposition and the crystallisation of amorphous silicon. The goal is to fabricate cells by depositing the μ-Si on cheap ceramic or foil substrates, or on glass.

8.8.2.   *Microcrystalline silicon solar cell design*

Figure 8.1 illustrates the basic problem with using silicon in thin film solar cells. As in any polycrystalline material, the diffusion length in μ-Si is effectively limited by the grain size, and this places a limit on the cell thickness for effective carrier collection of a few microns or tens of microns. However, the low absorption coefficient means that a layer of μ-Si a few microns thick harvests less than half of the available photons (Fig. 8.1). Below we mention two basic approaches to improving performance. A 2 μm thick μ-Si silicon solar cell on glass has been reported with to have an efficiency of over 10% [Yamamoto, 1998].

- Light trapping in thin film Si solar cells

  Relatively thin cells can be made from polycrystalline silicon if light trapping techniques are used to increase the optical path length inside the cell. Texturing front and back surfaces increases the optical depth by a factor of around 20, which means that only a few tens of microns of μ-Si are needed to absorb most of the incident light. This relaxes the need for a long diffusion length, and allows higher base doping, which increases V_{oc}. The physics of light trapping will be discussed in Chapter 9.

- Parallel multijunction thin film silicon solar cells

  An alternative concept in thin film silicon is the parallel multijunction solar cell, where the cell is composed of consecutive micron-thick layers of p and n type microcrystalline Si. Layers of similar polarity are connected together to give a set of p–n junctions connected in parallel. Because the layers are of similar thickness to the minority carrier diffusion length, the probability is high that a photogenerated minority carrier will reach and cross a p–n junction. Majority carriers then diffuse laterally towards the contacts. The cells perform better than single junction μ-Si cell of the same thickness because the fraction of minority carrier lost by recombination is smaller.

8.9. Summary

Thin film materials based on polycrystalline or amorphous semiconductors are being developed for lower cost photovoltaics. Advantages include the possibilities of deposition over large areas on cheap substrates (such as glass, plastic or foil) and the potential for fabricating integrally connected modules during film deposition. The primary disadvatange is lower power conversion efficiency resulting from poorer charge transport properties in the defective materials. Suitable materials should have high optical absorption. The leading materials are amorphous silicon (a-Si), and the polycrystalline semiconductors cadmium telluride (CdTe) and copper indium gallium diselenide (CIGS).

Amorphous silicon offers strong light absorption compared to crystalline silicon and may be deposited at lower temperatures. As an amorphous semiconductor, it suffers from 'dangling bond' defects which make doping difficult, and intra band tail states which reduce mobility by charge trapping. Passivation with hydrogen reduces the defect density sufficiently that useful devices may be made, but the efficiency of doping is low and lifetimes in doped material are very short. Poor transport in doped a-Si mean that solar cells are made in p–i–n rather than p–n structures. In p–i–n structures, collection losses may result from either neutralisation of electric field or series resistance in the i-region. Higher efficiencies may be obtained using tandem structures, which reduce the thickness of individual i regions, or heterojunctions, which reduce collection losses in the emitter by introducing a window. The dispersive transport characteristics and series resistance of a-Si lead to light intensity dependent transport and current voltage characteristics. A long-standing problem is light induced degradation or Staebler-Wronski effect, which causes a 30% reduction in cell performance over the first months of operation.

Polycrystalline semiconductors are characterised by grain boundary effects. Defects at grain boundaries in doped material introduce space charge regions where majority carrier mobility is reduced and minority carrier recombination is enhanced. Saturation of traps reduce grain boundary recombination at high light intensities. The effect of grain boundaries on mobility and lifetime can be described with a model based on the depletion approximation. In the best polycrystalline photovoltaic materials, CdTe and CIGS, grain boundary activity is minimised during materials growth, and efficient p–n junction devices can be made.

For both CdTe and CIGS, cells are prepared as n–p heterojunctions with an n-type CdS emitter in order to minimise QE losses in the emitter. The hetero-interface introduces additional defects which act as recombination centres and reduce the open circuit voltage. In CIGS-CdS structures the heterojunction also introduces a barrier in the conduction band impeding electron collection. The main challenges are to improve transport properties of the bulk materials; improve the quality of the heterojunction; improve contact with substrate and reduce losses due to surface recombination in the emitter. Efficiencies of over 16% and 18% have been achieved in CdTe and CIGS, respectively, in n–p structures on glass.

Microcrystalline silicon is attractive for thin film photovoltaics, but suffers from low optical absorption. Designs incorporating light trapping structures offer to increase the QE for micron-thick films and achieve improved transport properties by oparating at high carrier density.

References

M.D. Archer and R.D. Hill (eds.), *Clean Electricity from Photovoltaics* (London: Imperial College Press, 2001).

R.H. Bube, *Photovoltaic Materials* (London: Imperial College Press, 1998).

H.C. Card and E.S. Yang, "Electronic processes at grain boundaries under optical illumination", *IEEE Trans. Electron Devices* **ED-24**, 397–402 (1977).

M.A. Contreras *et al.*, "Progress toward 20% efficiency in Cu(In,Ga)Se polycrystalline thin-film solar cell", *Progr. Photovoltaics* **7**, 311–316 (1999).

H. Fritzsche, "Density of states in noncrystalline solids", in *Physical Properties of Amorphous Materials*, eds. D. Adler *et al.* (New York: Plenum, 1985).

M. A. Green, K. Emery, D. L. King, S. Igari and W. Warta, "Solar cell efficiency tables", *Progr. Photovoltaics* **9**, 287–293 (2001).

S.S. Hegedus, "Current-voltage analysis of a-Si and a-SiGe solar cells including voltage dependent photocurrent collection", *Prog. Photovoltaics Res. Appl.* **5**, 151–168 (1997).

P.T. Landsberg and M.S. Abrahams, "Effects of surface states and of excitation on barrier heights in a simple model of a grain boundary", *J. Appl. Phys.* **55**, 4284–4293 (1984).

J. Merten *et al.*, "Improved equivalent circuit and analytical model for amorphous silicon solar cells and modules", *IEEE Trans. Electron Devices* **45**, 423–429 (1998).

K. Mitchell, A.L. Fahrenbruch and R.H. Bube, "Photovoltaic determination of optical absorption coefficient in CdTe", *J. Appl. Phys.* **48**, 829–830 (1977).

H.J. Moller, *Semiconductors for Solar Cells* (Artech House, 1983).

P. de Pauw *et al.*, "On the injection level dependence of the minority-carrier lifetime in defected silicon substrates", *Solid State Electron.* **27**, 573 (1984).

U. Rau and H.W. Schock, "Cu(In,Ga)Se$_2$ solar cells", in *Clean Electricity from Photovoltaics*, eds. M.D. Archer and R.D. Hill (London: Imperial College Press, 2001).

J.Y.W. Seto, "The electrical properties of polycrystalline silicon films", *J. Appl. Phys.* **46**, 5247–5254 (1975).

W. Spear and P. LeComber, "Electronic properties of substitutionally doped amorphous Si and Ge", *Phil. Mag.* **33**, 935 (1976).

T. Tiedje, "Time resolved charge transport in hydrogenated amorphous silicon", in *Physics of Hydrogenated Amorphous Silicon*, eds. J.D. Joannapoulos and G. Lucovsky, 261–300 (Springer, 1984).

J.R. Tuttle, "The effect of composition on the optical properties of Cu$_2$InSe$_2$ thin films", *Conf. Record. 19th IEEE Photovoltaic Specialists Conf.*, 1494–1495 (1987).

K. Yamamoto *et al.*, "Thin film poly-si solar cell on glass substrate fabricated at low temperature", *MRS Spring Meeting*, April, 1998, San Francisco.

Chapter 9

Managing Light

9.1. Introduction

This chapter is about the effect of increasing photon flux density within the solar cell. The main reason for doing this is that by increasing the photon flux per unit volume, we decrease the volume of cell material which is needed to harness a certain amount of light. This decreases the *cost* per unit of energy converted, since the volume of cell material is the main determinant of cost. Another reason is that cells may perform more efficiently under increased photon flux levels.

We start by reviewing how light is utilised in a photovoltaic cell. Consider a semiconductor layer of active width w, supported by a substrate of a similar semiconductor, as illustrated in Fig. 9.1. Light normally incident on the cell passes through encapsulating materials before meeting the cell surface, where it may strike semiconductor or cell contacts. Some fraction, R, of the light will be reflected from the semiconductor surface, contacts and outer layers. Of the remainder, some will be absorbed in the active

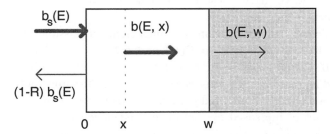

Fig. 9.1. Attenuation, transmission and reflection of incident radiation in a semiconductor layer of active width w. $b_s(E)$ is the incident photon flux density normal to the surface, and $b(E, x)$ is the transmitted photon flux density at depth x below the surface.

layers and some will pass straight through into the substrate. Within the cell, the flux attenuates so that at a depth x the flux density of photons of energy E is given by

$$b(E, x) = (1 - R(E))b_s(E)e^{-\int_0^x \alpha(E, x')dx'} \qquad (9.1)$$

where $b_s(E)$ is the incident photon flux density normal to the surface, $R(E)$ is the reflectivity of the surface, and $\alpha(E, x)$ is the absorption coefficient of the semiconductor at x. Assuming that all absorbed photons result in band gap excitation, electron-hole pairs are generated at a rate

$$g(E, x) = \alpha(E, x)b(E, x) = (1 - R(E))\alpha(E, x)b_s(E)e^{-\int_0^x \alpha(E, x')dx'} \qquad (9.2)$$

per unit volume, in accordance with Eq. 4.25. For a cell of thickness w, a fraction

$$f_{abs} = 1 - R(E) - \frac{b(E, w)}{b_s(E)}$$

of incident photons are absorbed while a fraction

$$f_{trans} = \frac{b(E, w)}{b_s(E)}$$

pass through. For a uniform absorber the absorbed and unused fractions are

$$f_{abs} = (1 - R(E))(1 - e^{-\alpha(E)w}) \qquad (9.3)$$

and

$$f_{trans} = (1 - R(E))e^{-\alpha(E)w} \qquad (9.4)$$

respectively. The unabsorbed fraction is high whenever αw is small, for instance, for wavelengths close to the band gap and for indirect semiconductors such as silicon.

For efficient light utilisation, as much light as possible should be absorbed in the active layers, so $R(E)$ and $b(E, w)$ should be as small as possible. If $b(E, x)$ could be increased within the cell, the photogeneration rate would increase. This means we could achieve the same photogeneration rate with a smaller volume, reducing materials requirements.

The effect of increased $b(E, x)$ on cell efficiency is discussed below in Sec. 9.4. We will show that if the flux density is increased by a factor X, the cell current should increase, to a first approximation, in proportion to X while the cell voltage increases as $\log(X)$. This means that the power delivered by the cell, $J(V) \times V$, increases *faster* than X, so the *efficiency* increases under increased photon flux levels. Note that this is independent of the result that for some defective semiconductors, current collection can be improved at higher light intensities.

This chapter is organised as follows: first we review the role of photon flux in a solar cell and discuss the ways in which light management can improve the performance of a solar cell. Then we treat the following strategies: anti-reflection coatings, concentrator systems, light trapping structures and photon recycling systems. In each case the effect on device physics is briefly discussed.

9.2. Photon Flux: A Review and Overview of Light Management

Box 9.1. Understanding photon flux: some definitions

The spectral photon flux density $\beta(E, \mathbf{r}, \theta, \phi)$ is the number of photons of energy E which pass through unit area per unit solid angle per unit time. In general, β is defined by a direction (θ, ϕ) and is position as well as energy dependent. Conservation of photons requires that, for some volume $\mathcal{V}$ of material with surface S,

$$\int_S \beta(E, \mathbf{s}, \theta, \phi) d\Omega \cdot d\mathbf{S} = \int_\nu (g_{\mathrm{ph}}(E, \mathbf{r}) - u_{\mathrm{ph}}(E, \mathbf{r})) dV \qquad (9.5)$$

where $\mathbf{s}$ is a point on the surface, $g_{\mathrm{ph}}(E, \mathbf{r})$ is the photon emission rate and $u_{\mathrm{ph}}(E, \mathbf{r})$ the photon absorption rate per unit volume. We define $u_{\mathrm{ph}}(E, \mathbf{r})$ as the *net* absorption, that is, the difference between absorption and stimulated emission. The left hand integral in Eq. 9.5 is taken over solid angle Ω and points $\mathbf{s}$ on the surface S, while the right hand integral is taken over all points $\mathbf{r}$ within the volume $\mathcal{V}$. Using Gauss's law we find that at any point within $\mathcal{V}$

$$\int_\Omega \boldsymbol{\nabla} \cdot \beta(E, \mathbf{r}, \boldsymbol{\theta}, \boldsymbol{\phi}) d\Omega = g_{\mathrm{ph}}(E, \mathbf{r}) - u_{\mathrm{ph}}(E, \mathbf{r}). \qquad (9.6)$$

The rate of absorption for photons for a given flux direction is proportional to the magnitude of the photon flux, β, in that direction, and the emission rate proportional to the magnitude of emitted flux. Summing up contributions over solid angle, we find

$$u_{ph}(E, \mathbf{r}) = \int_\Omega \alpha(E, \mathbf{r})\beta(E, \mathbf{r}, \theta, \phi)d\Omega \tag{9.7}$$

where α is the absorption coefficient at $\mathbf{r}$, and

$$g_{ph}(E, \mathbf{r}) = \int_\Omega \varepsilon(E, \mathbf{r})\beta_e(E, \mathbf{r}, \theta, \phi)d\Omega \tag{9.8}$$

where ε is the emission coefficient and and β_e is the emitted flux density. Note that, in general, ε and α may be direction dependent. If they are direction independent, then u_{ph} and g_{ph} are proportional to the sum of magnitudes of the fluxes from different directions. Combining Eqs. 9.6 to 9.8 and resolving along the direction (θ, ϕ), we have

$$\nabla \cdot \beta = \varepsilon\beta_e - \alpha\beta$$

which can also be written as

$$\frac{d\beta}{dl} = \varepsilon\beta_e - \alpha\beta \tag{9.9}$$

where l is the lenth coordinate along (θ, ϕ) [Chandrasekhar, 1950].

In real materials the emitted flux, which arises from the radiative recombination of carriers, is usually negligible. That leaves us with the soluble linear equation

$$\frac{d\beta}{dl} + \alpha\beta = 0 \tag{9.10}$$

In the case where the incident flux is directed normal to the cell surface $(\theta = 0, \phi = 0)$, we set $\beta = b(E, x)$ and have the one dimensional equation

$$\frac{db}{dx} + \alpha b = 0 \tag{9.11}$$

which has the solution

$$b(E, x) = b(E, 0)e^{-\int_0^x \alpha(E, x')dx'} \tag{9.12}$$

usually known as the Beer–Lambert law. Henceforth, $b(E, x)$ is used to represent the flux travelling at depth x normal to the surface of a planar structure.

9.2.1. *Routes to higher photon flux*

An increased photon flux can be achieved in several ways, by *concentration* of the light, *trapping* of the light, or by exploitation of *photon recycling*. These are illustrated schematically in Fig. 9.2.

All of these processes aim to improve the efficiency of utilisation of photons in a given amount of photovoltaic material. They are useful in different conditions, for instance light trapping strategies are particularly important for weak absorbers and thin cells, while concentration is important in very efficient, crystalline cells.

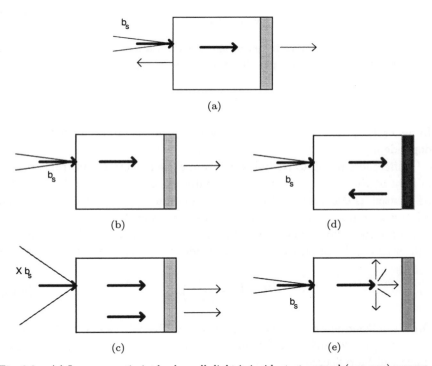

Fig. 9.2. (a) In a non optimised solar cell, light is incident at normal (one sun) concentration and light is lost by reflection and transmission. Photogeneration can be increased by (b) reducing reflection with an anti-reflection coating; (c) increasing the incident photon flux density by concentrating the light. Note that this increases the angle subtended by the light source; (d) confining the light by means of internal reflection (or light trapping) strategies; (e) reabsorption of photons produced by radiative recombination. Some of these strategies would normally be used together, *e.g.*, an AR coat would be used under concentration, and light trapping structures would be used to enhance photon recycling.

In this chapter we will look at the basic physics of each of the processes, how they are or might be achieved in practice, and how they affect the physics and performance of the photovoltaic cell.

9.3.   Minimising Reflection

9.3.1.   *Optical properties of semiconductors*

The optical properties of a solid are described by the dielectric constant, ε_s. ε_s is a complex quantity and obeys

$$\sqrt{\varepsilon_s} = n_s - i\kappa_s \tag{9.13}$$

where n_s is the refractive index of the material and the imaginary part κ_s is related to the absorption coefficient of the material through

$$\alpha = \frac{4\pi\kappa_s}{\lambda}. \tag{9.14}$$

In general, ε_s, n_s and κ_s are wavelength dependent, and may be direction dependent.

Differences in refractive index determine the reflection and transmission of light at the interface between two materials. For thin films, light should be treated as coherent, and Maxwell's equations can be solved to find the relative amplitudes of transmitted and reflected waves. In the simplest case, that of a plane boundary between materials of refractive index n_0 and n_s, light striking the interface at normal incidence is reflected with probability

$$R = \left(\frac{n_0 - n_s}{n_0 + n_s}\right)^2. \tag{9.15}$$

For a semiconductor, n_s is typically 3–4 at visible wavelengths (Table 9.1), so that some 30–40% of light normally incident on the surface from free space will be reflected.

At oblique incidence the reflectivity is angle and polarisation dependent. If the incident ray makes an angle θ_0 with the surface normal, and the transmitted ray an angle θ_s with the surface normal *inside* the semiconductor then

$$R = \left(\frac{\eta_0 - \eta_s}{\eta_0 + \eta_s}\right)^2 \tag{9.16}$$

Table 9.1. Refractive indices of selected semiconductors and AR coat materials. From various sources including Pankove [Pankove, 1971].

Material	Refractive index (at ca. 1.5 eV)
Si	3.44
Ge	4.00
GaAs	3.6
$Al_{0.8}Ga_{0.2}As$	3.2
AlAs	3.0
Tl_2O_5	2.1
ZnO	2.02
$Si_3 N_4$	1.97
SiO_2	1.46

where

$$\eta_s = n_s \sec \theta_s \tag{9.17}$$

for p polarised light (with the electric field vector *in* the plane of incidence) and

$$\eta_s = n_s \cos \theta_s \tag{9.18}$$

and for s-polarised light (electric field vector *normal* to the plane of incidence). θ_0 and θ_s are related through Snell's law

$$n_0 \sin \theta_0 = n_s \sin \theta_s \,.$$

For unpolarised light, considered as an equal mix of s and p polarisations, the net reflectivity generally increases with angle (for the s polarised component it increases while for the p component it has a minimum), approaching one at large angles. The minimum reflectivity of a semiconductor surface to unpolarised light occurs at normal incidence and is given by Eq. 9.15. At 30–40%, this is unacceptably high for efficient photovoltaic energy conversion.

Snell's law also means that light travelling *within* the semiconductor towards the surface at an angle greater than the critical angle, θ_c, is internally reflected. For an air-semiconductor interface the critical angle is given by

$$\theta_c = \sin^{-1} \left(\frac{n_0}{n_s} \right). \tag{9.19}$$

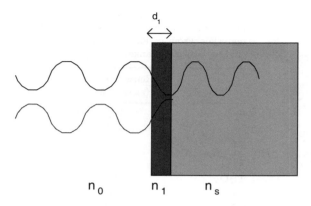

Fig. 9.3.   Single layer anti-reflection coating. When the thickness of the dielectric layer is equal to one quarter wavelength, the incident and reflected waves interfere destructively to cancel out the reflected ray at the outer surface.

For the air-silicon interface $\theta_c = 16.9°$. In the presence of an optical coating n_0 in Eq. 9.19 is replaced by the refractive index of the coating and θ_c is increased. This phenomenon of total internal reflection is exploited in light trapping structures, which are discussed in Sec. 9.6.

9.3.2.   *Anti-reflection coatings*

Reflectivity of the air-semiconductor interface can be reduced using an *anti-reflection (AR) coating*. An AR coat is a thin film of a dielectric with refractive index n_1 intermediate between those of the semiconductor (n_s) and free space (n_0). By considering forward and backward travelling waves in each medium it can be shown [Macleod, 2001] that the reflectivity of the film for light of wavelength λ is given by

$$R = \frac{(\eta_0 - \eta_s)^2 + (\eta_0\eta_s/\eta_1 - \eta_1)^2 \tan^2 \delta_1}{(\eta_0 + \eta_s)^2 + (\eta_0\eta_s/\eta_1 + \eta_1) \tan^2 \delta_1} \qquad (9.20)$$

where δ_1 is the phase shift in the film,

$$\delta_1 = 2\pi\eta_1 d_1 \cos\theta_1/\lambda, \qquad (9.21)$$

θ_1 is the angle between the light ray and the normal within the film, and d_1 is the film thickness. R clearly has its minimum value when $\delta_1 = \pi/2$. For normal incidence, this first happens when d_1 is equal to a quarter wavelength in the thin film material. (In those conditions the waves reflected from the front and rear interface of the thin film are out of phase and interfere destructively.) R *vanishes* when it is also true that

$$n_1 = \sqrt{n_0 n_s}\,. \tag{9.22}$$

So by coating our semiconductor with a thin layer of a medium with refractive index $\sqrt{n_s}$, we can reduce the reflectivity to zero at some particular wavelength, λ_0. Since the solar spectrum is broad, this wavelength should be chosen to lie towards the middle of the range of wavelengths which can be usefully absorbed for that semiconductor, and this determines the AR coat thickness. Close to λ_0, R increases with wavelength approximately like $(\frac{\Delta\lambda}{\lambda_0})^2$, and at wavelengths where the phase shift becomes a multiple of π, R reaches its maximum value equal to the natural reflectivity of the uncoated interface. This means that an AR coat optimised at one visible wavelength may be quite highly reflecting at others. AR coats on silicon solar cells are usually optimised for red light, where solar irradiance is strong, and become reflective in the blue. For this reason silicon solar cells often appear violet or blue. Similarly, the reflectivity of the AR coat depends on angle, and increases at wide angles. This means that AR coats are of limited use in diffuse light or for non-planar surfaces. Figure 9.4 shows how an AR coat which is tuned for normal incidence at a certain wavelength becomes reflecting at other wavelengths and wide angles.

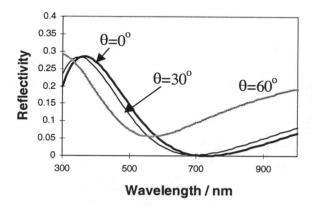

Fig. 9.4. Calculated reflectivity of single layer AR coat as function of wavelength at different angles of incidence $\theta = 0°$, $30°$ and $60°$. Notice how the minimum of reflectivity lifts and shifts to shorter wavelengths at wider angles of incidence. Calculated for a 100 nm layer of refractive index $n_1 = \sqrt{3.3}$ on top of a semiconductor of refractive index $n_s = 3.3$.

In the above treatment of AR coats, we neglect the absorption of the optical materials. This is usually a good approximation since, for most semiconductors, the absorption visible wavelength is sufficiently weak that $\kappa_s \ll n_s$. For reflection from an absorbing medium, n_s should properly be replaced by the complex refractive index $n_s - i\kappa_s$ in Eqs. 9.17–9.20 and R calculated with complex arithmetic. The *incident* medium should of course be non-absorbing.

According to Eq. 9.22, the ideal AR coat material for silicon should have a refractive index of around 1.84, and for GaAs, 1.90. Good materials are silicon nitride which has $n_1 = 1.97$ and tantalum oxide with $n_1 = 2.1$. These give relectivities of less than 1% at normal incidence at the optimum wavelength.

Improved reflectivity over a band of wavelengths can be achieved with two or more thin films. The greater the number of layers, the greater the range of wavelengths over which the reflectivity can be minimised. Multiple layers are not usually practical for solar cells, given their cost and the sensitivity to angle of incidence, but double layers are used on some high efficiency cells. Layers should be deposited so that refractive index increases consecutively from air (n_0) to first coat (n_1) to second coat (n_2) to semiconductor (n_s). Reflectivity vanishes when both films have quarter-wave

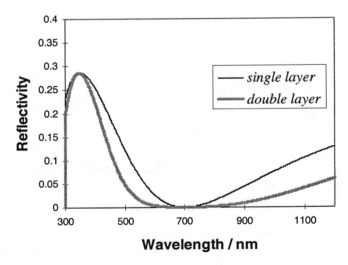

Fig. 9.5. Calculated reflectivity of double layer AR coat, optimised for 700 nm. The reflectivity of a single layer is presented for comparison.

thickness and

$$\left(\frac{n_2}{n_1}\right)^2 = \frac{n_s}{n_0}. \tag{9.23}$$

This condition allows a little more flexibility in choice of materials than the single layer case. In some cell designs, where a wide band gap window layer is present for improved carrier collection, a double layer AR coat can be made simply by choosing an optical coating with refractive index n_1 satisfying Eq. 9.23 where n_2 is the refractive index of the window layer, and choosing appropriate thicknesses. This has been done with GaAs devices having a high aluminium content AlGaAs window.

Regarding solar cell device physics, the effect of anti-reflection coatings is simply to reduce R, and increase the flux and generation rates, as defined in Eqs. 9.1 and 9.2.

9.4. Concentration

9.4.1. *Limits to concentration*

In photovoltaic *concentration* systems, light is collected over a large area and focused on to a cell surface of smaller area. This increases the incident flux density at the cell surface by a concentration factor, X, which is approximately the ratio of the collection and cell areas. The main objective of this strategy is to reduce the cost of the photovoltaic conversion system. Concentrating optics are much cheaper per unit area than photovoltaic cells and the cells convert the concentrated and unconcentrated light with approximately the same efficiency. To a first approximation, the concentrator cell delivers the power of X identical cells in unconcentrated light.

According to geometrical optics, concentrating the incident light is equivalent to expanding the angular range subtended by the source. Concentrating sunlight by a factor X using a spherical concentrator increases the angular range to θ_X such and enhances the total (angle-integrated) incident flux by a factor

$$X = \frac{\int_0^{2\pi} \int_0^{\theta_X} d\Omega}{\int_0^{2\pi} \int_0^{\theta_s} d\Omega} = \frac{\sin^2 \theta_X}{\sin^2 \theta_{\text{sun}}} \tag{9.24}$$

where θ_{sun} is the half angle subtended by the sun, $0.26°$. It is assumed that the intensity of the concentrated light is uniform over the angular range. Because of refraction of light, most of the light entering the semiconductor is directed close to the normal, and so the flux normal to the surface is enhanced approximately by the factor X relative to the unconcentrated case. For normal incidence,

$$b(E, x) = (1 - R(E))Xb_s(E)e^{-\int_0^x \alpha(E,x')\mathrm{d}x'} . \tag{9.25}$$

To a first approximation, the generation rate, as given in Eq. 9.2 increases by the same factor.

The geometrical factor F_s, introduced in Chapter 2 to relate the absorbed light to the angular range of the source, changes to

$$F_X = \pi \sin^2 \theta_X = X \sin^2 \theta_{\mathrm{sun}} . \tag{9.26}$$

The maximum theoretical concentration occurs when $\theta_X = 90°$, *i.e.*, $X = 46050$. For a concentrator with cylindrical rather than spherical symmetry the maximum concentration is lower: $X = 1/\sin \theta_{\mathrm{sun}}$.

9.4.2. *Practical concentrators*

In practice, concentration is limited by the quality of the optics, by tracking errors in working photovoltaic systems, and by the practical difficulty of coupling very highly concentrated light, with its very wide angular range, into the cell. Practical concentration ratios of tens to thousands of suns have been achieved using reflective optics or Fresnel lenses.

Note that imaging concentration is only possible when incident light rays are parallel and the concentrator is pointing at the sun. This means that only *direct* sunlight can be concentrated. *Diffuse* sunlight, which has been scattered by the atmosphere and composes some 15% of the standard

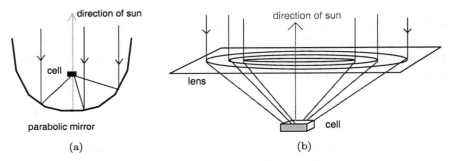

Fig. 9.6.   Concentrator system based on (a) parabolic reflector (b) Fresnel lens. The cell is placed at the focus in either case.

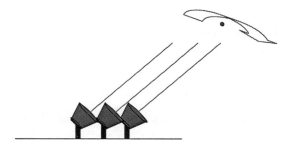

Fig. 9.7. **Tracking system.** The concentrator must be mounted to follow the direction of the sun on a daily **and** seasonal cycle.

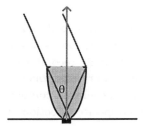

Fig. 9.8. Compound parabolic concentrator.

AM1.5 spectrum, is incident on the concentrator from wide angles and cannot be coupled into the cell. Even in direct sunlight, the concentrator must be made to follow the sun during its daily and seasonal trajectories, and this is achieved with *tracking* systems. Concentrator systems are most cost effective with high quality cells where the saving on photovoltaic materials is greatest, and in sunny climates where the direct solar irradiation is largest. Tracking systems are essential for high efficiency but add to the cost and complexity of the system.

An alternative to imaging concentrators are non-imaging, low concentration systems which do not require tracking. An example is the compound parabolic concentrator [Welford, 1990]. The cell receives light with the angular range θ, where θ is the half angle of the paraboloid, and achieves concentration ratios of $4/\sin^2 \theta$, or $4n^2/\sin^2 \theta$ if filled with a medium of refractive index n.

Another approach to non-imaging concentration is the use of fluorescent concentrators. A refractive material filled with fluorescent particles re-emits photons on irradiation. The emitted light can be directed on to the cell exploiting total internal reflection, and the spectral range of the re-emitted light can be controlled [Goetzberger, 1977].

9.5.   Effects of Concentration on Device Physics

We will distinguish cells which operate under concentration in *low injection* conditions, where photogenerated carrier density is smaller than the equilibrium majority carrier density, from those which operate in *high injection* conditions, where the photogenerated density is greater.

9.5.1.   *Low injection*

In low injection, a *p–n* concentrator cell can be treated essentially with the standard theory for a *p–n* junction as developed in Chapter 6. In low injection conditions, minority carrier recombination rates are linear and the current densities resolve into a dark component J_{dark} which depends on V and a bias independent, light generated component J_{sc} which depends on flux density and acts in the opposite direction. Thus, from Eq. 1.4

$$J(V) = J_{\text{sc}} - J_{\text{dark}}(V)$$

where the dark current typically has the diode-like form so that

$$J(V) = J_{\text{sc}} - J_0(e^{qV/mk_BT} - 1). \tag{9.27}$$

Now if the incident flux b_s is increased by a factor X through concentration of the light, J_{sc} will increase to a first approximation by the same factor, *i.e.*, $J_{\text{sc}}(Xb_s) \approx XJ_{\text{sc}}(b_s)$, but leave the dark current, for a given bias, unaffected. This means that at any given bias below V_{oc} the *net* current increases with X. Moreover, a higher bias is needed to achieve the open-circuit condition where dark current cancels photocurrent. Replacing J_{sc} with XJ_{sc} in Eq. 9.27 we see that for a diode of ideality factor m, the open circuit voltage becomes

$$V_{\text{oc}}(X) = \frac{mk_BT}{q} \ln\left(\frac{XJ_{\text{sc}}}{J_0} + 1\right) \approx V_{\text{oc}}(1) + \frac{mk_BT}{q} \ln X \tag{9.28}$$

where $V_{\text{oc}}(1)$ represents the open circuit voltage under unconcentrated light (1 sun). Thus V_{oc} increases logarithmically with X, and if the cell fill factor remains the same then the power delivered by the cell should increase by a

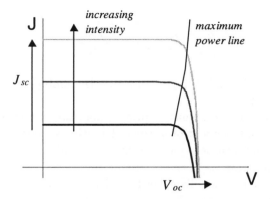

Fig. 9.9. Current–voltage curve at different intensities. J_{sc} increases linearly with light intensity and V_{oc} increases like the logarithm of intensity. The result is that the maximum power point moves to higher voltages, and the efficiency increases.

factor

$$X \left(1 + \frac{mk_{B}T}{qV_{oc}(1)} \ln X \right),$$

and the efficiency by a factor

$$\left(1 + \frac{mk_{B}T}{qV_{oc}(1)} \ln X \right).$$

Figure 9.9 shows how the operating point of the cell moves to higher voltage as X increases.

In practice, however, the dark current *is* affected by concentration through (i) increasing carrier densities leading to high injection conditions, and (ii) increasing temperature, both leading to deviations from this behaviour. These effects are discussed in the following sections.

9.5.2. High injection

Concentrator cells may be designed to operate in *high injection* conditions. That means that the photogenerated carrier densities are large compared to the doping density in the base, and both carrier types must be included in the transport equations. Generally n and p will be different and electric fields due to carrier density gradients must be taken into account. In the limit where $n = p$ we showed in Sec. 8.4, for p–i–n structures, that the transport equation for either carrier reduces to the diffusion equation-like form,

$$\frac{d^2n}{dx^2} - \frac{(n - n_0)}{L_a^2} + \frac{g(E, x)}{D_a} = 0 \qquad (8.9^*)$$

where $D_a = \frac{D_n D_p}{D_n + D_p}$, $L_a^2 = D_a \tau_a$, and τ_a is the ambipolar carrier lifetime, given by Eq. 8.12.

If the dominant recombination process is *linear* with excess carrier density, then L_n is independent of n, the differential equation (8.9) is linear in n and the solution can be separated into independent light dependent and bias dependent terms, as in the treatment in Sec. 6.5, *i.e.*, superposition is valid. This is the case for Shockley Read Hall recombination, even at high injection. Letting $n = p$ in Eq. 4.79 for the SRH recombination rate, and assuming $n \gg n_t$, and $p \gg p_t$, we find that

$$U_{SRH} \approx \frac{n}{\tau_n + \tau_p}. \qquad (9.29)$$

U_{SRH} is still proportional to n but the carrier lifetime has been increased. Physically this is due to the relative scarcity of majority carriers. In low injection, the capture of a minority carrier in a trap state is immediately followed by the capture of a majority carrier, simply because of the abundance of majority carriers, and the rate determining step was the minority carrier capture. Now both hole capture *and* electron capture are rate determining.

If SRH recombination is dominant at all injection levels, superposition should hold and the arguments given above for the variation of V_{oc} and cell power with concentration should be valid. Indeed, J_{sc} may increase even faster than X, on account of the longer carrier lifetime at high injection.

Radiative and Auger recombination, however, become non linear with n in high injection. From Eq. 4.57 it is clear that the radiative recombination rate will vary like

$$U_{rad} = Bn^2$$

when $n = p$, implying a lifetime which varies like $\tau \propto n^{-1}$. Similarly, band-to-band Auger recombination (Eq. 4.66) varies like

$$U_{Aug} = An^3$$

implying that $\tau \propto n^{-2}$. Therefore at injection levels where radiative and Auger processes are dominant, superposition becomes invalid, the dark current term increases with generation rate and V_{oc} rises more slowly than predicted by Eq. 9.28.

The different recombination regimes may be distinguished as a change in the intensity dependence of V_{oc}. If the separation in electron-hole Fermi levels is equal to qV and $n = p$, then from the law of mass action we have

$$n = n_i e^{qV/2k_BT}$$

and so we expect recombination to vary like $\sim e^{qV/2k_BT}$ for SRH process, like $\sim e^{qV/k_BT}$ for radiative processes and like $\sim e^{3qV/2k_BT}$ for Auger. Consequently we expect V_{oc} to vary like $\frac{2k_BT}{q} \ln X$, $\frac{k_BT}{q} \ln X$ and $\frac{2k_BT}{3q} \ln X$, respectively, in the three cases. In the case of silicon, Auger recombination places a limit on the cell performance under concentration.

9.5.3. *Limits to efficiency under concentration*

In Sec. 2.5 we showed how to calculate the efficiency of a solar cell in the radiative limit, where the only loss process is radiative recombination. The net flux absorbed in the cell is the difference between an incident flux term which is proportional to F_s, and an emitted flux term which is proportional to F_e, the geometrical factor for emitted light. For a flat plate cell in a standard sun $F_s/F_e = 2.16 \times 10^{-5}$, because of the small angular width of the sun. Increasing F_s to F_X by concentrating the light (Eq. 9.25) increases the *absorbed* relative to *emitted* fluxes at any value of the chemical potential (see Eq. 2.16). This increases the product of net flux and chemical potential, and hence the limiting efficiency of the solar cell. Maximum concentration is achieved when the angle of acceptance θ_X reaches 90°, and $F_X = F_e = \pi$ radians, so that the cell accepts and emits light over the same angular range. This leads to a maximum efficiency of over 40% if the temperature is unaffected. Another way of achieving the condition that $F_e = F_s$ is to *restrict* the angular range for *emission*. This idea is developed in Chapter 10.

In practice, concentration systems increase the maximum efficiency of high quality Si and GaAs *cells* by a few percent. However, when the optical losses in the concentrating system and the wastage of diffuse light are taken into account, concentrator modules actually perform no more efficiently than one sun modules in AM1.5. Concentrators are of interest primarily for the reduced cost through reduced cell area, and not for improved efficiency.

Changing spectral intensity also changes the optimum band gap. Figure 9.10 shows the limiting efficiency of an ideal photoconverter, calculated as a function of band gap according to Sec. 2.4–2.5, for an AM1.5 spectrum with $X = 1$ and $X = 1000$ ($F_X = 1000F_s$). It is clear that the maximum efficiency point shifts to lower band gap as concentration increases.

Note that this calculation does not include Auger recombination, which is, like radiative recombination, an unavoidable process, nor the effects of temperature.

9.5.4. *Temperature*

High concentration raises the *temperature* of the cell. At higher temperature the intrinsic carrier population

$$n_i^2 \sim e^{-E_g/k_B T}$$

increases, an effect which is enhanced by the shrinkage of the band gap E_g with increasing temperature. Increasing n_i increases the dark current. The reduced band gap increases the photocurrent, but only slightly compared to the effect on J_{dark}, so the increasing dark current dominates. This opposes the effect of increasing efficiency under concentration with the result that optimum performance is achieved at some intermediate concentration level. GaAs has a slightly better temperature coefficient of efficiency than silicon and finds its optimum at a higher concentration level.

9.5.5. *Series resistance*

By increasing the current density, concentration increases the potential drop, JAR_s, due to *series resistance* in the cell. This is a problem particularly in the emitter layer where lateral current densities are high. Concentrator cells must therefore be designed differently, with different metallisation patterns and ideally with higher emitter doping than one-sun cells. Emitter doping levels are limited by recombination and cannot be increased indefinitely, nor can the contacted area become too large, so compromises must be found. Minimisation of series resistance losses is a major objective in concentrator cell design.

9.5.6. *Concentrator cell design*

Concentrator cells should be of very high purity, since high efficiency is a priority, and because the cell is a smaller fraction of the system cost. The concentrator cell should have very low bulk recombination losses, so high quality crystals should be used, and good surfaces. Light trapping is helpful in reducing the cell thickness and thereby reducing resistive losses due to carrier diffusion in the high injection limit. More lightly doped material can be used for the active layers than for a one sun cell because high carrier densities mean higher conductivity, and also because lower doping produces better quality material. Examples of concentrator cells are the PERL cell, developed at UNSW, ($\eta = 26\%$ in 20 suns) which operates in low injection conditions, and the rear point contact cell, developed at Stanford, which is based on an undoped substrate and operates in high injection [Sinton, 1986; Swanson, 1990]. Buried contacts which combine low resistivity with low shading offer to improve series resistance in future designs.

As noted above, higher emitter current densities mean that series resistance minimisation through contact and emitter design is important. Contact grids should have thicker, denser fingers and often are designed with symmetry to match the image of the concentrated light. They are normally designed to give maximum performance under certain concentration levels, and do not as a rule give the best performance under one-sun conditions. The higher degree of metallisation needed increases losses by reflectivity and shading (to 10–20%), but this can be overcome by encapsulating the cell beneath a prismatic lens which refracts the light onto the bare cell surface and away from the contact fingers.

9.5.7. *Concentrator cell materials*

Figure 9.10 shows that the optimum band gap for photovoltaic energy conversion is reduced under concentration. This would suggest that silicon, with $E_g = 1.1$ eV, is a better concentrator cell material than GaAs with $E_g = 1.4$ eV. However, GaAs has two important advantages: (i) GaAs enjoys a better temperature coefficient of efficiency than silicon and (ii) Auger recombination is stronger in silicon because of its lower band gap and limits the theoretical efficiency to 36% at a concentration of 100 suns with perfect light trapping. Auger recombination is much less important in GaAs and efficiency continues to rise with concentration up to several hundred suns. At 1000 suns the theoretical efficiency is 36–37% [Sinton, 1995; Klausmeier-Brown, 1995].

No simple direct semiconductors exist with band gaps in the range 1.1–1.3 eV, where the theoretical efficiency under concentration is high. Some ternary alloys, such as $In_x Ga_{1-x} As$, have suitable band gaps but their material quality is inferior. To date, GaAs remains the best material for high concentration with a record efficiency of 27.6% in AM 1.5 under 255 suns [Vernon, 1991]. The best silicon cell is 26.8% at 96 suns [Verlinden, 1995].

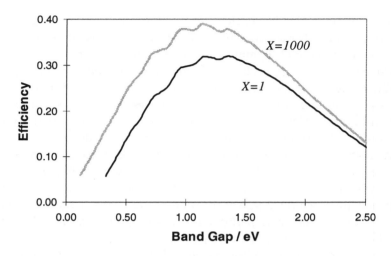

Fig. 9.10. Limiting efficiency versus band gap for an ideal solar cell in AM 1.5 with $X = 1000$ and $X = 1$.

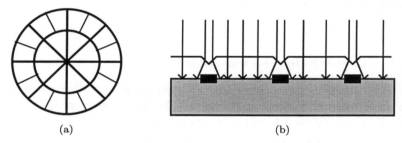

(a)   (b)

Fig. 9.11. (a) Contact grid for concentrator cell with circular symmetry; (b) Prismatic lens refracts light on to cell surface away from contacts.

9.6. Light Confinement

9.6.1. *Light paths and ray tracing*

In contrast to anti-reflection coats, which increase the fraction of photons admitted to the cell, and concentration which increases the incident flux, light confinement techniques increase the path length of photons *inside* the cell, once admitted. Increasing the path length increases the probability of photogeneration per incident photon, particularly when the absorption coefficient is small, increasing the absorbed fraction. Light trapping is normally considered in the regime of geometrical optics where structures are large compared to the coherence length of the light, and light rays with different history do not interfere. This is a good approximation in silicon where cells are hundreds of microns thick. In micron scale structures, light should be treated as coherent and interference becomes important. In such systems, classical ray tracing approaches are not valid, and the photogeneration rate must be found from the gradient of the Poynting vector.

Light trapping involves the scattering and reflection of light rays within the cell so that they travel at wider angles to the surface normal. The problem can be considered in terms of the *paths* taken by light rays inside the cell. An admitted ray may travel in one of a number of paths, P_i, which are defined by the cell geometry. At some point within the cell volume, the total flux is given the sum of the contributions along the different paths. The relevant quantity for photogeneration is the light absorption rate, from Eq. 9.7,

$$u_{ph}(E, \mathbf{r}) = \alpha(E, \mathbf{r}) \sum_{P_i} \beta_i(E, \mathbf{r}, \theta_i, \phi) \qquad (9.30)$$

where the integral over solid angle is converted to a sum over the paths, and β_i represents the photon flux at $\mathbf{r}$ from the direction (θ_i, ϕ_i). Equation 9.30 provides the photogeneration rate at $\mathbf{r}$ and should replace Eq. 9.2 in the calculation of photogeneration. However, because of the different history of the rays it is difficult to determine the β_i at a general point, and it is easier to find the total absorbed flux by considering absorption along the length of each ray path, using an approach known as 'ray tracing'. Thus,

$$f_{abs} = \frac{(1 - R(E))}{b_s(E)} \left\{ 1 - R(E) - \sum_i \beta_i(E, 0) \exp\left(-\alpha(E) \int_{P_i} dl_i \right) \right\} \qquad (9.31)$$

where l_i is a length co-ordinate along the path P_i, and the integration is taken over the ray path, from entry to exit. $\beta_i(E, 0)$ is the flux of the ith ray at the start of its path inside the cell — effectively a weighting factor for the path — and satisfies

$$\sum_{P_i} \beta_i(E, 0) = (1 - R(E))b_s(E). \qquad (9.32)$$

For clarity, we have assumed a position independent absorption coefficient.

Clearly a longer ray path will lead to a higher attenuation of the ray and a higher f_{abs}. Different light trapping schemes can be evaluated by comparing the average path length

$$\langle l \rangle = \sum_{P_i} \beta_i(E,0) \int_i dl_i \bigg/ \sum_{P_i} \beta_i(E,0) \qquad (9.33)$$

with the width of the cell, w. Note that the path length is dependent only on geometry and reflection and transmission probabilities, and does not contain the probability of photon absorption.

9.6.2. *Mirrors*

The simplest light trapping scheme is to introduce an optical mirror at the rear surface of the cell, either by metallising the rear cell surface or by growing the active layers on top of a Bragg stack. The mirror typically reflects over 95% of rays striking the rear surface. Rays which subsequently reach the front, semiconductor-air, surface are likely to pass through since the reflectivity of that interface must be small for efficient light capture. So the rear mirror effectively doubles the path length of the light. For an ideal mirror (with $R = 1$) and ideal front surface (with $R = 0$) the path length is $2w$.

Greater path length enhancements can be achieved by exploiting total internal reflection at the front surface. This is not possible for a cell with parallel planar surfaces since, by symmetry, reflected rays must approach the front surface at the same angle at which they were admitted, and consequently encounter the same reflectivity, which should be small. For light trapping, rays should approach the front surface at a wider angle than the angle at which they were admitted. This can be achieved by tilting the front or the rear surface, or by scattering the light within the cell. If light rays can be manipulated so that they approach the surface at an angle $\theta_s > \theta_c$

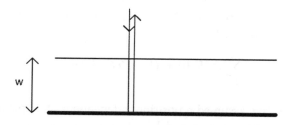

Fig. 9.12.   Double path length in metallised cell.

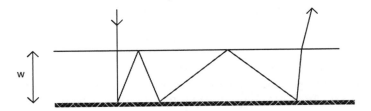

Fig. 9.13. Ideal randomising rear surface. Light rays striking the rear surface are reflected in a random direction. On the first two reflections, the illustrated ray is reflected at an angle larger than θ_c. On the third bounce, it is reflected with $\theta < \theta_c$ and escapes.

they will be unable to escape from the cell and be reflected back for another double pass within the cell.

9.6.3. *Randomising surfaces*

One approach is to use a roughened surface which scatters, or *randomises*, the light upon reflection, changing the value of θ_s. Here we consider a randomising rear surface although in principle either the front or the rear surface, or both, can be randomising. A perfectly randomising surface, called a *Lambertian* surface, scatters reflected rays with spherical symmetry so that all angles in the hemisphere pointing into the cell are equally likely. Assuming normal incidence, a light ray enters and crosses the cell and is scattered at the rear surface. The mean path length for the reflected ray to cross back to the front surface is twice the cell width:

$$\langle l \rangle = \frac{w}{\langle \cos \theta_s \rangle} = 2w \, . \tag{9.34}$$

This doubling of the optical depth of the cell illustrates the advantage of randomising the light even before light confinement has occurred. Now, since only those rays with $\theta_s > \theta_c$ can escape, only $\frac{1}{n_s^2}$ of the rays striking the front surface will leave the cell; the rest will be reflected to make a second pass, of length $2w$, across the cell and be scattered again at the rear surface. Adding up the path lengths of the rays according to the number of times they are scattered before leaving the cell we find

$$\langle l \rangle = \frac{1}{n_s^2} \times 2w + \frac{1}{n_s^2} \times \left(1 - \frac{1}{n_s^2}\right) \times 6w + \frac{1}{n_s^2} \times \left(1 - \frac{1}{n_s^2}\right)^2 \times 10w + \cdots$$

where it is assumed that the path length is w for the first pass (normal incidence) and the final pass (only small θ_s can escape) but $2w$ for all intermediate passes. Summing the series to infinite order,

$$\langle l \rangle = (4n_s^2 - 2)w \approx 4n_s^2 w. \tag{9.35}$$

This ideal path length enhancement, $4n_s^2$, is around 50 for silicon and GaAs, which shows the potential of light trapping systems [Lush, 1991].

If the rear surface is imperfectly reflecting then a fraction of the light is lost through the rear surface at each scattering event, and the mean path length is reduced. The effect can be considerable because of the large number of reflection events (around n_s^2) at that surface.

Although useful for indicating the potential of light trapping systems, perfectly randomising surfaces are not easily realised. Practical techniques usually depend on the geometrical-optical properties of more regular, textured surfaces.

9.6.4. *Textured surfaces*

The simplest system is where one surface is tilted relative to the other. Consider a rear surface tilted at an angle θ_{tilt} relative to the planar front surface, as in Fig. 9.14. When $\theta_{\text{tilt}} > \frac{1}{2}\theta_c$, normally incident rays will be reflected from the rear surface at an angle greater than θ_c, and be totally reflected at the front. If a ray strikes the same portion of the rear surface on the second pass, it will be reflected at an even wider angle, and trapped again. For uniform cell width, both positive and negative tilt angles must be present, so that trapped rays will eventually be reflected at narrower angles and escape. If the positive and negative tilt angles are equal, then each ray makes a multiple of four passes across the cell. The worst case, four passes (a 'double bounce'), is illustrated in Fig. 9.14. The overall mean path length depends upon the size and tilt angles of the texture, but it can

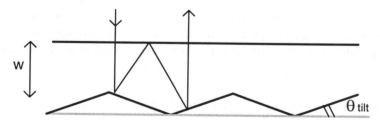

Fig. 9.14.   Light trapping structure with a tilted rear surface, illustrating a 'double bounce' light path.

immediately be stated that

$$\langle l \rangle \geq 2(1 + \sec 2\theta_{\text{tilt}})w \gtrsim 4w. \tag{9.36}$$

Although higher tilt angles lead to higher path length enhancements for the shortest pass, shallower angles allow longer textured faces, and a higher probability of multiple passes.

Texturing the front surface in the same way leads to similar path length enhancements. Normally incident rays striking the textured surface are refracted into the cell at an angle

$$\gamma = \theta_{\text{tilt}} - \sin^{-1}\left(\frac{\sin\theta_{\text{tilt}}}{n_s}\right) \tag{9.37}$$

which increases the path length from side to side of the cell. If the texturing is symmetrical, then most of those rays will strike the front surface again on a face which has opposite tilt to the face of entry, and leave the cell. So, at the minimum, only a double pass is available and

$$\langle l \rangle \geq 2w \sec\gamma > 2w. \tag{9.38}$$

However, if the texturing is asymmetrical, then by choosing the tilt angles carefully, four passes of the light can be achieved, as shown in Fig. 9.15.

As a general trend, the lower the degree of symmetry, the greater the degree of light trapping. If the surface is textured in two directions, with pyramids rather than grooves, then rays are scattered in three dimensions rather than two and light trapping is improved. Calculation of the mean path length for any texturing pattern is complicated and requires the average of ray paths over the different points of incidence by *ray tracing*. However, it is easy to see that for a surface textured with pyramids the probability of a light ray striking the surface at the same angle at which it was admitted — the condition for escape — is smaller than for a surface

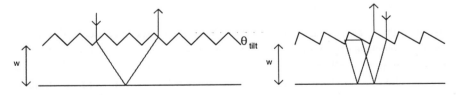

Fig. 9.15. (a) A simple reflection in a structure with symmetrical surface texture. Notice how the surface texture extends the path length from front to rear of the cell. (b) A four pass path in a cell with asymmetric surface texture.

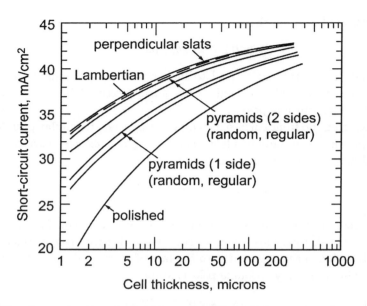

Fig. 9.16. J_{sc} as a function of cell thickness, for different light trapping schemes. [Swanson, 1990.]

textured only with grooves. For a *regular* array of pyramids, light confinement is better if the pyramids are offset from each other. *Random* arrays of pyramids, which have a very low degree of symmetry, trap light very well and are comparable with the ideal randomising surface. Inverted regular pyramids, as used in the PERL cell (Sec. 7.4) perform slightly better than upright ones because of an increased probability of the escaping light re-entering the cell. Figure 9.16 compares the light trapping performance of a range of different surfaces.

9.6.5. *Practical schemes*

Front surface texturing in monocrystalline materials may be achieved by using a selective etch which acts in preferred directions to expose particular crystal planes. In silicon a basic etch can be used on the (100) surface to to expose the (111) crystal faces. The size of surface pyramids can be defined before etching by using a mask.

Surface texturing is less important for polycrystalline materials. Depending on the method of film deposition, the surface is already likely to be rough and therefore have a lower reflectivity than single crystal. Chemical etches have been used in polycrystalline silicon to create 'honeycomb' and other structures which enhance light trapping.

Textured rear surfaces can be prepared by growing the active layers of the cell on patterened substrates. These can be prepared by etching; or by epitaxial methods, and the active layers then grown epitaxially on top. The size of features on textured surfaces are typically in the μm range, which is acceptable for silicon where cells are at least tens of microns thick, but problematic for highly absorbing materials like GaAs. An alternative is to grow the active layers over a blazed grating [Heine, 1995] which achieves the light scattering without increasing cell width.

Approximately randomising front surfaces have been prepared on silicon using random pyramidal surface texture [Green, 1995], and alternatively with a thin layer of porous silicon on top of the active layers. The porous silicon coating can be produced by electrochemically etching the silicon surface. Nanometre sized crystallites and pores in the porous silicon scatter the light to achieve wider angles of entry [Stalmans, 1998].

Reflective rear surfaces are prepared by metallising the rear of the cell with aluminium or gold. Such semiconductor-metal interfaces have reflectivities of over 95%. Reflectivity at that surface can be improved if a thin layer of dielectric — typically SiO_2 — is inserted between semiconductor and metal. Rays incident at wide angles are then reflected by total internal reflection. For epitaxially grown crystals, an alternative is to grow the active layers on top of a Bragg stack, which has very high reflectivity over a band of wavelengths.

Actual path length enhancements of over 10, rather smaller than the maximum of $4n_s^2$, have been achieved in silicon cells with roughened surfaces and with pyramidal texturing. Note that this is considerably smaller than the flux enhancements available through concentration; unlike concentrators, light trapping cells operate in relatively low injection conditions.

We have not discussed the application of resonant cavities and photonic band gap to photovoltaics. These are useful for strong confinement of particular wavelengths or bands of wavelengths, and may be of interest for photovoltaic effects with narrow band light sources, but are not expected to be beneficial for broad band sources.

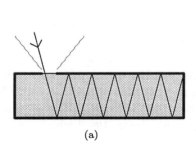

(a)   (b)

Fig. 9.17. (a) Highly confining cell structure with mirrored surface; (b) A reflective cavity containing arrays of cells (the 'photovoltaic eye').

9.6.6. *Light confining structures: restricted acceptance areas and external cavities*

The degree of light trapping can be increased if the cell is enclosed in a reflective cavity with restricted aperture for acceptance of light. Two examples are illustrated in Fig. 9.17. In the first, the entire surface of the cell is coated with a reflective material, apart from a small aperture which is open to incident light. Rays admitted through this aperture will be reflected many times before they escape and have a path length many times the width of the cell. In the second example, many cells are mounted in an reflective cavity which has a small aperture for acceptance of light. In this case, the path length of rays inside the cavity is long, and rays which are not absorbed in one cell are likely to be absorbed by another, so the path length of rays in the assembly of cells is long compared to the width of a cell [Minano, 1990].

In both cases the aperture area, A_{aperture}, is smaller than the geometrical surface area of the cell or cells, A_{cell}, and although the optical path length is increased, say by $\frac{\langle l \rangle}{w}$, the photon flux inside the cell material is increased by a much smaller factor, approximately $\frac{\langle l \rangle}{w} \times \frac{A_{\mathrm{aperture}}}{A_{\mathrm{cell}}}$, which is comparable to the enhancement available with a rear mirror. Such cavity structures are really only useful in conjunction with concentrating systems, where light is concentrated on to the aperture to increase the photon flux admitted and consequently increase the photon flux density inside the cell material.

9.6.7. *Effects of light trapping on device physics*

Incorporating light trapping can be compared to increasing the cell thickness. To see this, compare a standard cell of width w without rear reflector with a light trapping cell of width $W = \frac{w}{X}$ and a mean optical path length of $\langle l \rangle = WX$. The extended path length compensates for the reduced cell thickness so that, to a first approximation, the fraction of light absorbed is the same in either case.

Now we can make the following observations.

- Bulk recombination losses are reduced. The thickness of the light trapping cell is smaller compared to the minority carrier diffusion length, so a smaller fraction of the photogenerated carriers should be lost by minority carrier recombination than in the standard cell.
- Rear surface recombination losses may be increased. The point generation rate is made up of contributions from reflected and incident rays and is more uniform than for the standard cell. This means that generation will be significant near the rear surface, and the fraction of carriers available for recombination there is higher. Very good quality rear surfaces are therefore needed, with low recombination velocities as well as very high optical reflectivities.
- Bulk resistivity losses are reduced. Because the same amount of generation is occurring in a smaller volume, the photogenerated carrier densities are higher, and the conductivity is increased.
- The probability Auger recombination is increased, again because of higher carrier densities.

The procedure for calculating the carrier and current densities in the light trapping structure is similar to that used for a standard cell, such as the p–n junction example in Chapter 6. The main modification is that the generation rate, given by Eq. 9.2, must be replaced by a function, such as Eq. 9.30, which incorporates the contributions from rays which have been reflected from the front and rear surfaces. Provided that the carrier densities are still low enough for the Shockley Read Hall recombination rate to vary linearly with carrier density in the neutral regions, *i.e.*, to satisfy low injection conditions, then the net current still decouples into a bias dependent dark current and a light dependent photocurrent. Comparing the same two cells as above, we see that the photocurrent should be similar in the two cases, although it may be slightly higher in the light trapping cell because of reduced bulk recombination losses. The dark current at any given applied bias will be smaller for the light trapping structure because it is thinner. Consequently the net photocurrent, $J(V) = J_{sc} - J_{dark}(V)$, will be increased.

The low injection approximation is likely to be valid for *practical* light trapping structures where path length enhancements are seldom greater than 10. In some very lightly doped device structures light trapping may push the device into the high injection regime, where recombination becomes non-linear. This regime is discussed above in Sec. 9.5.2.

Comparison of light confinement with concentration

Let's compare a concentrator cell of width w and concentration factor X with a light trapping cell of the same width and path length enhancement $\langle l \rangle = wX$. The cells have the same volume and the same average photon flux density. If only bulk recombination is important, then the cells should perform equivalently since there will be the same probability of a recombination event per generation event in either case. However, if surface recombination *is* important, then the light trapping cell will perform less well. Because of the higher number of reflections per generation event, more carriers are likely to be generated near the rear surface, and surface recombination will therefore be greater.

The maximum theoretical power density enhancement available with light trapping is the same as that available with concentration: $4n^2/\sin^2 \theta_{sun}$ where θ_{sun} is the half angle subtended by the sun. In practice, however, it is harder to achieve high enhancement ratios with light trapping than to achieve high concentration factors.

9.7.   Photon Recycling

9.7.1.   *Theory of photon recycling*

Photon recycling (PR) is the reabsorption of photons emitted by radiative recombination inside the cell. Until now, it has been assumed that the photon generation term in Eq. 9.6 is negligible. This is a good approximation in materials where non-radiative processes dominate. Nevertheless, we know from Chapters 2 and 4 that in a semiconductor away from equilibrium, radiative recombination of electrons and hole is unavoidable, and so there is in fact a non-zero photon generation term in the photon continuity equation, though it may be extremely small. Including this term, Eq. 9.9 should be used

$$\frac{d\beta}{dl} = \alpha\beta_e - \alpha\beta \tag{9.39}$$

where we have used the detailed balance result, that $\varepsilon = \alpha$. $\beta(E, \mathbf{r}, \theta, \phi)$ is, as usual, the component of the flux along (θ, ϕ), and β_e is the magnitude of emitted flux per solid angle, given by Eq. 2.11

$$\beta_e(E, \mathbf{r}, \theta, \phi) = \frac{2n_s^2}{h^3 c^2} \frac{E^2}{e^{(E-\Delta\mu)/kT_c} - 1} \tag{9.40}$$

where $\Delta\mu$ is the splitting in the quasi Fermi levels at $\mathbf{r}$ and n_s the refractive index. Thus, when photon recycling is included, the photon flux density at any point within the cell is increased by the spontaneous emission of photons from other points.

Let's look at the solution to the inhomogenous differential equation, Eq. 9.39. If we consider a system with planar symmetry, then β depends only on the depth x into the cell and we may convert the co-ordinate l to x. For light normally incident on to a planar cell, we define

$$l = x \sec\theta \tag{9.41}$$

where θ is the angle between the ray direction and normal to the cell surface.

In the absence of photon recycling, the first term on the right-hand side of Eq. 9.39 is zero and the solution for β is the Beer–Lambert form. This is the *externally generated* flux, and it has the form

$$\beta(x) = (1 - R)\beta_s e^{-\alpha x \sec\theta} = \beta(0)e^{-\alpha x \sec\theta} \quad \text{for } 0 \le \theta \le \theta_c \tag{9.42}$$

where β_s is the angularly resolved emission from the sun (Eq. 2.1) and α is assumed uniform. In the following, the parameter E has been dropped for clarity and β_e is expressed as a function of x, indicating the dependence on $\Delta\mu$ at x.

Now, when photon recycling is included, Eq. 9.39 has the general solution

$$\beta(x) = \beta(0)e^{-\alpha x \sec\theta} + \Delta\beta(x)$$

$$\Delta\beta(x) = \int_0^x \alpha\beta_e(x')e^{-\alpha(x-x')\sec\theta} \sec\theta dx' \tag{9.43}$$

which shows that the flux at any point inside the cell is increased by contributions emitted from all other points. To find the carrier generation rate at x we sum the flux over all ray directions, (Eq. 9.8)

$$g(E,x) = \int \alpha\beta(x)d\Omega = 2\pi \int \alpha\beta(x) \quad d\cos\theta, \tag{9.44}$$

whence

$$g(E,x) = (1-R)\alpha b_s e^{-\alpha x}$$

$$+ \alpha \int_{-1}^{1}\int_{0}^{x} \alpha\beta_e(x')e^{-\alpha(x-x')\sec\theta}\sec\theta dx'd\cos\theta \tag{9.45}$$

where we have used the fact that $\cos\theta_c \sim 1$ at a semiconductor-air interface, and the relationship between $\beta(0)$ and b_s. Renaming the terms we have

$$g(E,x) = g_{\text{ext}}(E,x) + g_{\text{in}}(E,x) \tag{9.46}$$

where the first term in Eq. 9.46 is solely due to *external* generation, and the second to *internal* generation. It is clear from Eqs. 9.40 and 9.45 that $g_{\text{in}}(E,x)$ depends on the electron and hole densities at x' through the quasi Fermi level separation at that point, $\Delta\mu(x')$, and hence is coupled to the differential equations for electron density, hole density and Poisson's equation. In general, the integral Eq. 9.44 for $g(E,x)$ needs to be added to the differential system and solved numerically. The relationship between $g_{\text{in}}(E,x)$ and $\Delta\mu(x)$ depends upon the geometry of the cell and degree of light trapping, as is illustrated in the example below.

Some approximations can be made in the calculation of photon recycling effects [Parrott, 1993; Balenzategui, 1995; Marti, 1997].

(i) Most simply, $\Delta\mu(x)$ is set to a constant, whence $\beta_e(x)$ is constant. This is a good approximation for very thin cells, where $g_{\text{ext}}(E,x)$ is roughly constant;

(ii) $\Delta\mu(x)$ is set to the value it would have if photon recycling was ignored, and $\beta_e(x)$ is worked out from that;

(iii) Iterating (ii), $\Delta\mu(x)$ is recalculated from the approximate result for $\beta_e(x)$, and that is used to improve the solution for $\Delta\mu(x)$. Successive iterations should yield a solution for $\Delta\mu(x)$ which is stable to some desired tolerance. Using this method, it was estimated that photon recycling could increase the efficiency of a GaAs cell with a rear mirror and high radiative efficiency by 1.3%, and the V_{oc} by tens of mV [Balenzategui, 1995]. Another way to look at photon recycling is an extension of the radiative *lifetime* leading to reduced radiative recombination rate.

9.7.2. Practical schemes

Unlike other strategies discussed above, photon recycling is an intrinsic process which happens in all illuminated semiconductors. It is relevant only in materials with high radiative efficiency — direct band gap semiconductors such as GaAs — and can be exploited only in devices which are already limited by radiative recombination. Since all practical solar cells are limited by non-radiative processes, managing photon recycling to achieve higher efficiency is still a theoretical objective.

Photon recycling clearly increases with average ray path length and therefore with the degree of light trapping. As shown in the example below, the effect is stronger in a cell with good rear reflectivity. As material quality improves, photon recycling will need to be included as a factor in the design of highly efficient solar cells. Note that the effect is included by default in the calculation of limiting efficiency in Chapters 2 and 10.

Box 9.2. Photon recycling in a cell with rear mirror

$g_{in}(x)$ depends upon the geometry of the cell and degree of light trapping. Here we consider the case of a planar cell which has a perfectly reflecting rear surface and a front surface with zero reflectivity within the critical angle. The additional flux at a depth x is given by the second term in Eq. 9.45. The angular integral contains contributions from the following three types of trajectory:

- $\theta < \theta_c$ Ray path direct from front surface. No reflections. (Left-hand ray in Fig. 9.18.)
- $\theta_c < \theta < (\pi - \theta_c)$ Ray reflected more than once from front or rear surfaces. (Centre rays in Fig. 9.18.)

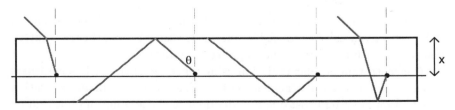

Fig. 9.18.   Photon recycling in a cell with a perfect rear reflector. The figure illustrates the four types of rays which need to be included.

- $\pi - \theta_c < \theta < \pi$ Ray reflected once at rear surface. (Right hand ray in Fig. 9.18).

From Eq. 9.43 we have for $\theta < \theta_c$,

$$\Delta\beta(x) = \int_0^x \alpha\beta_e(x')e^{-\alpha(x-x')\sec\theta}\sec\theta dx' \qquad (9.47a)$$

for $\theta_c < \theta < (\pi - \theta_c)$,

$$\Delta\beta(x) = \int_0^x \alpha\beta_e(x')e^{-\alpha(x-x')\sec\theta}\sec\theta dx'$$
$$+ \left(\frac{1}{1-e^{-2\alpha w\sec\theta}}\right)\int_0^{2w}\alpha\beta_e(x')e^{-\alpha(x-x')\sec\theta}\sec\theta dx' \qquad (9.47b)$$

and for $\pi - \theta_c < \theta < \pi$,

$$\Delta\beta(x) = \int_x^w \alpha\beta_e(x')e^{-\alpha(x-x')\sec\theta}\sec\theta dx'$$
$$+ e^{-\alpha(w-x)\sec\theta}\int_0^w \alpha\beta_e(x')e^{-\alpha(w-x')\sec\theta}\sec\theta dx' . \qquad (9.47c)$$

Equation 9.47 can be solved iteratively to yield a solution for $\Delta\beta(x)$ and corresponding $\Delta\mu(x)$. Notice that potentially the largest contribution comes from the second term of the second result, which accounts for multiple reflections of internally reflected rays. The higher the degree of light trapping, the greater the effect on $\Delta\beta$.

9.8. Summary

Photocurrent generation in a solar cell can be increased if the photon flux density within the cell can be increased. The total photon flux available for photogeneration consists of the flux density incident on the cell surface, less the fractions which are reflected and transmitted without being absorbed. The incident flux density can be increased by concentrating the light; reflectivity can be reduced by surface treatments; and the transmitted component reduced using light trapping structures.

Concentration of solar radiation using reflective optics or Fresnel lenses increases the flux density incident on the cell surface. The main motivation is reduction in system cost due to reductions in the amount of photovoltaic material needed. In theory, concentration ratios of over 46,000 are possible but in practice the detrimental effects of increased temperature and series resistance losses limit concentration ratios to a few hundred suns. Exploiting concentration effectively requires mechanical tracking systems which follow the path of the sun, and this adds to the cost. Lower levels of concentration can be achieved with passive, non-tracking optical concentrators. Concentration increases the limiting efficiency to over 40%, and efficiencies of up to 28% have been achieved in practice. It is primarily of interest for high efficiency solar cells. Under high concentration levels, cells work in the 'high injection' limit where carrier densities are high, recombination through defects is suppressed and Auger and radiative recombination processes become more important.

The reflectivity of a semiconductor at visible wavelengths is typically over 30%. This can be reduced to a few percent using an anti-reflection coating, which is a thin film of dielectric material with refractive index smaller than the cell material. Anti-reflection coats are normally optimised for a certain wavelength, and are less effective at different wavelengths and off-normal incidence. They are routinely used in crystalline solar cells. Reflectivity may also be reduced by texturing the surface, and this technique is used in crystalline and polycrystalline materials.

Light trapping structures confine the light by exploiting internal reflection and scattering. Strategies include reflective mirrors at the rear of the cell, textured front and rear surfaces. The effective path length of a light ray within the cell is increased by multiple reflections or by refraction at wide angles to the normal. The photon flux density at any point within the cell is increased by contributions from internally reflected rays, resulting in higher effective optical depth. Light trapping is usually treated with geometrical optics and quantified by the enhancement of the optical path length compared to the cell without light trapping. Path length enhancements of around 50 are theoretically possible, and enhancements of around 10 have been reported in practice. Light trapping is routinely used in silicon and is of strong interest for thin film silicon photovoltaics.

In materials with high radiative efficiencies, the useful flux density can be increased by reabsorption of emitted photons or 'photon recycling'. The photon flux at a point is increased by the emitted flux from other points due to the light-induced quasi Fermi level separation, and the quasi Fermi level separation is increased by the enhanced light absorption. Photon recycling effects are useful only in materials where radiative recombination is an important loss mechanism and they are larger within light trapping structures. They are not yet exploited within any practical cell structures.

References

J.L. Balenzategui, A. Marti and G.L. Araujo, "Numerical modelling of photon recycling effects in solar cells using PC1D", *Proc. 13th European Photovoltaic Solar Energy Conf.*, 1223–1226 (1995).

S. Chandrasekhar, *Radiative Transfer* (Oxford University Press, 1950).

A. Goetzberger and W. Greubel, "Solar energy conversion with fluorescent concentrators", *Appl. Phys. Letts.* **14**, 123 (1977).

M.A. Green, *Silicon Solar Cells: Advanced Principles and Practice* (Sydney: Centre for Photovoltaic Engineering, 1995).

C. Heine and R.H. Morf, "Submicrometer gratings for solar-energy applications", *Appl. Opt.* **34**, 2476 (1995).

M.E. Klausmeier-Brown, "Status, prospects and economics of terrestrial single junction GaAs concentrator cells", in *Solar Cells and their Applications*, ed. L.D. Partain (Wiley, 1995).

G. Lush and M. Lundstrom, "Thin film approaches to high efficiency III-V cells", *Solar Cells* **30**, 337–344 (1991).

H.A. Macleod, *Thin Film Optical Filters* (Bristol: Institite of Physics, 2001).

A. Marti, J.L. Balenzategui and R.F. Reyna, "Photon recycling and Shockley's diode equation", *J. Appl. Phys.* **82**, 4067–4075 (1997).

J.C. Minano, "Optical confinement in photovoltaics", in *Physical Limitations to Photovoltaic Energy Conversion*, eds. A. Luque and G.L. Araujo (Adam Hilger, 1990).

J.I. Pankove, *Optical Processes in Semiconductors* (Englewood Cliffs: Prentice Hall, 1971).

J.E. Parrott, "Radiative recombination and photon recycling in photovoltaic solar cells", *Sol. En. Materials Sol. Cells* **30**, 221–231 (1993).

R. Sinton *et al.*, "27.5% silicon concentrator solar cells", *Electron Device Letts.* **EDL-7**, 567 (1986).

R.A. Sinton, "Terrestrial silicon concentrator cells", in *Solar Cells and their Applications*, ed. L.D. Partain (Wiley, 1995).

L. Stalmans *et al.*, "Porous silicon in crystalline silicon solar cells: A review and the effect on the internal quantum efficiency", *Progr. Photovoltaics* **6**, 233–246 (1998).

R.M. Swanson and R.A. Sinton, "High efficiency silicon solar cells", *Advances in Solar Energy* **6**, 427–484 (1990).

P.J. Verlinden *et al.*, "A 26.8% efficient concentrator point-contact solar cell", *Proc. 13th European Photovoltaic Solar Energy Conf.*, 1582–1585 (1995).

S.M. Vernon *et al.*, "High-efficiency concentrator cells from GaAs on Si", *Conf. Record, 22nd IEEE Photovoltaic Specialists Conf.*, 353–357 (1991).

W.T. Welford, "Limits to concentration by passive optical systems", in *Physical Limitations to Photovoltaic Energy Conversion*, eds. A. Luque and G.L. Araujo (Adam Hilger, 1990).

Chapter 10

Over the Limit: Strategies for High Efficiency

10.1. Introduction

A photovoltaic cell operates by converting light energy into electricity. It is a quantum energy conversion process, whereby packets of light energy, *photons*, are consumed to deliver units of electrical charge (*electrons*) to an external circuit where they do electrical work. The power conversion *efficiency* of a solar cell is a measure of the amount of work done per photon. Increasing the efficiency is essentially a matter of increasing the ratio of work extracted to photon energy supplied. This chapter is concerned with ways — both practical and theoretical — of increasing that ratio.

In Chapter 2 we saw how the limiting efficiency of a simple, single junction solar cell is calculated by considering detailed balance. This limit, of 33% in a standard air mass 1.5 (AM 1.5) spectrum, follows from a number of assumptions. Important amongst these are the assumptions that exactly one electron-hole pair is generated by each absorbed photon, that electrons and holes relax to form populations in thermal equilibrium with the lattice, and that all the light is absorbed in a junction of single band gap. Given these constraints, the majority of the sun's energy is lost by (i) failure to capture the energy of photons with energy (E) smaller than the band gap (E_g), and (ii) thermal dissipation of kinetic energy of carriers generated with $E \gg E_g$, as shown in Fig. 10.1. Processes (i) and (ii) account for about 23% and 33% of the sun's energy, respectively. The ratio of work out to photon energy in is greatest for photons with energy close to the band gap; elsewhere it is zero (for the low energy photons where $E < E_g$), or poor (for $E \gg E_g$). Figure 2.9 compares the power available at different photon energies with the incident power spectrum. Losses result from the poor match of the single band gap of the solar cell with the broad spectrum of solar radiation.

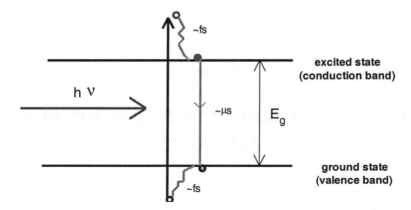

Fig. 10.1. Schematic energy band diagram for a single band gap solar cell. Photons of energy $h\nu$ greater than the band gap, E_g, are capable of promoting electrons to the excited state (the conduction band in a semiconductor). The excess energy is quickly lost as heat as the excited carriers thermalise to the band edges. Relaxation across the band gap is much slower, and in a photovoltaic device the thermalised carriers can be extracted before they finally recombine.

Chapters 3–9 were all concerned with devices and materials which operate within the limits presented in Chapter 2. In this chapter we discuss ways in which the theoretical limiting efficiency may be increased. A separate route to increasing the limiting efficiency, by concentration of solar radiation to expand the acceptance solid angle, has already been discussed in Chapter 9.

In the following sections we first review the calculation of the detailed balance limiting efficiency and underlying assumptions. Then we consider a number of strategies for increasing the amount of work done per photon by relaxing the assumptions underlying the detailed balance limit. Briefly, these strategies are concerned with:

- increasing the number of band gaps, to utilise different photon energies more efficiently (tandem and other multi-band solar cells);
- reducing the dissipation of thermal energy by photogenerated carriers ('Hot carrier' solar cells); and
- increasing the number of electron-hole pairs per photon (impact ionisation solar cells).

Apart from tandem solar cells, the strategies are mainly theoretical and would require materials with properties which have not yet been realised.

However, all are based on physical observations which suggest that the proposed phenomena *may* be exploited in practice.

10.2. How Much is Out There? Thermodynamic Limits to Efficiency

We start by calculating the maximum work available from a photoconverter considering only thermodynamic arguments. Essentially, the solar cell is a cool body which is radiatively coupled to its environment and operates by absorbing short wavelength radiation from a hot body (the sun) at temperature T_s and allowing some of the absorbed energy to be extracted as work. To satisfy detailed balance the device must also emit longer wavelength radiation characteristic of its own temperature T_c. In the simplest picture, both sun and cell absorb and emit light like black bodies at their respective temperatures. A black body at temperature T emits an amount of energy given by σT^4 per unit time and surface area, where σ_S is Stefan's constant. If the incident radiation is fully concentrated and the only loss process is spontaneous emission by the cell, then the net energy flux density received by the solar cell is given by

$$\sigma_S T_s^4 - \sigma_S T_c^4 .$$

Work can be extracted from the hot cell by means of a heat engine. Least energy is lost if the work is extracted isoentropically by means of a Carnot engine. Then the work available is

$$W = (\sigma_S T_s^4 - \sigma_S T_c^4)\left(1 - \frac{T_a}{T_c}\right) \tag{10.1}$$

where the final factor represents the Carnot factor for extraction of work from the cell at constant entropy. The power conversion efficiency is the ratio of extracted to incident power,

$$\eta = \frac{W}{\sigma_S T_s^4} = \left(1 - \left(\frac{T_c}{T_a}\right)^4\right)\left(1 - \frac{T_a}{T_c}\right). \tag{10.2}$$

For a sun temperature of 5760K and cell temperature of 300K, this has a maximum of 85% at a cell temperature of 2470K. In this limit, all photons are absorbed and the maximum amount of work is extracted per photon. Thermal dissipation is not considered. This limit is identical to the limiting efficiency of a solar thermal converter.

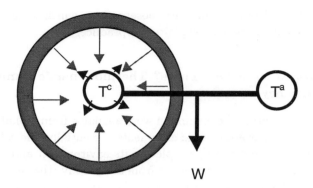

Fig. 10.2. Schematic of an ideal solar energy converter. The incident radiation is fully concentrated so that the cell absorbs and emits radiation in all directions. The cell, at T_c, is coupled to the environment, at T_a, through a heat engine which extracts work, W.

10.3.   Detailed Balance Limit to Efficiency, Reviewed

In Chapter 2, we calculated the limiting efficiency of a two level, or single band gap, photoconverter. Here we review that calculation and and introduce some of the concepts which will be used later in this chapter. A more detailed treatment of detailed balance as applied to solar cells may be found elsewhere [Araujo, 1994; de Vos, 1990].

The photoconverter has a band gap E_g, below which photons are not absorbed, and above which photons are absorbed to generate electron-hole pairs in thermal equilibrium with the lattice. It is usually assumed that the cell has planar geometry, and is able to receive and emit radiation into a hemishpere. For unconcentrated sunlight, the cell receives solar radiation through a small azimuthal range $0 < \theta < \theta_{\text{sun}}$ and ambient radiation through the remaining range $\theta_{\text{sun}} < \theta < \pi/2$. Assuming that the only loss process is radiative recombination, then the current output is equal to q times the difference between the photon fluxes which are absorbed and emitted through the cell surface, where q is the charge on the electron. Formally,

$$I = q \left\{ \begin{array}{l} \displaystyle\iint a(E,\mathbf{s},\theta,\phi)\beta_{\text{s}}(E,\mathbf{s},\theta,\phi)dE\,d\mathbf{S}\cdot d\Omega \\[2mm] \displaystyle+\iint a(E,\mathbf{s},\theta,\phi)\beta_{\text{a}}(E,\mathbf{s},\theta,\phi)dE\,d\mathbf{S}\cdot d\Omega \\[2mm] \displaystyle-\iint a(E,\mathbf{s},\theta,\phi)\beta_{\text{e}}(E,\mathbf{s},\theta,\phi)dE\,d\mathbf{S}\cdot d\Omega \end{array} \right\} \qquad (10.3)$$

where $a(E, \mathbf{s}, \theta, \phi)$ is the *absorptivity*, or probability of photon absorption, for light of energy E approaching along the direction (θ, ϕ) at a point $\mathbf{s}$ on the surface. $\varepsilon(E, \mathbf{s}, \theta, \phi)$ is the *emissivity*, or probability of photon emission, at that point. β, β_a and β_e are the photon fluxes incident from the sun, incident from the ambient, and emitted from the cell, respectively. Photon flux is defined as the number of photons per unit solid angle passing through unit area in unit time. The angular integrals for β_s and β_a should be taken over the appropriate range, as shown in Fig. 10.3.

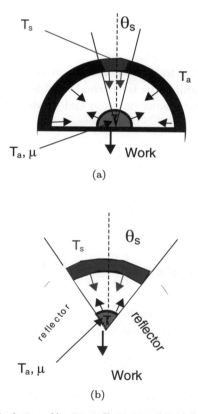

(a)

(b)

Fig. 10.3. (a) In the calculation of limiting efficiency, a flat plate cell at T_a receives light from sun within an angular range $\theta < \theta_{\text{sun}}$. The cell emits into a hemisphere. Emission into the rear hemisphere is prevented by a reflector. The chemical potential of the cell is maintained at a positive value, $\Delta\mu$, by the incident radiation. For concentrated light, the angular range of the sun is increased and with it the efficiency of the cell; (b) When the angular range of emission is restricted to be the same as the range of incidence, the performance is equivalent to the case of full concentration.

It is assumed for simplicity that both sun and ambient radiate like black bodies, producing isotropic photon fluxes described by the Planck distribution. So for the black body sun at T_s, we have

$$\beta_s = \frac{2}{h^3 c^2} \frac{E^2}{e^{E/k_B T_s} - 1} \tag{10.4}$$

over the range $0 < \theta < \theta_{sun}$. For the ambient environment at temperature T_a we have

$$\beta_a = \frac{2}{h^3 c^2} \frac{E^2}{e^{E/k_B T_s} - 1} \tag{10.5}$$

over the range $\theta_{sun} < \theta < \pi/2$.

The cell is assumed to possess a uniform chemical potential, $\Delta\mu$, on account of the light, and to be in thermal equilibrium with the ambient at T_a. This leads to spontaneous emission and produces an isotropic emitted flux of

$$\beta_e = \frac{2}{h^3 c^2} \frac{E^2}{e^{(E-\Delta\mu)/k_B T_s} - 1} \tag{10.6}$$

for $0 < \theta < \pi/2$.

For maximum performance the cell should absorb all incident photons above band gap. Therefore we assume

$$a(E) = \left\{ \begin{array}{ll} 1 & E \geq E_g \\ 0 & E < E_g \end{array} \right\}. \tag{10.7}$$

Two further simplifications result from the following. First, we assume that the chemical potential is constant and equal to q times the applied bias,

$$\Delta\mu = qV. \tag{10.8}$$

This implies that no potential is lost between light absorption and charge collection, *i.e.*, charge transport is lossless and mobilities are infinite. Second, we use the generalised detailed balance result that $\varepsilon(E) = a(E)$. This has been shown to be the case when when $\Delta\mu$ is uniform and equal to qV [Araujo, 1994].

For the geometry of a flat plate photoconverter of area A with a perfect rear reflector, the angular integrals can now be evaluated easily. Using the limits for θ given above we find for the current, I, at a bias V

$$I(V) = qA \left\{ \frac{2F_s}{h^3c^2} \int \frac{E^2}{e^{E/k_BT_s} - 1} dE + \frac{2(F_a - F_s)}{h^3c^2} \int \frac{E^2}{e^{E/k_BT_a} - 1} dE \right.$$

$$\left. - \frac{2F_a}{h^3c^2} \int \frac{E^2}{e^{(E-qV)/k_BT_a} - 1} dE \right\} \tag{10.9}$$

where the constants

$$F_a = \pi \tag{10.10}$$

and

$$F_s = \pi f_s \tag{10.11}$$

where $f_s = \sin^2 \theta_s = 2.16 \times 10^{-5}$ represents the angular range of the sun. If the light were concentrated by a factor X then

$$F_s = \pi X f_s . \tag{10.12}$$

Full concentration is reached when $X = 1/f_s$ and $F_s = \pi$.

It will be useful to define the function

$$N(E_{\min}, E_{\max}, T, \Delta\mu) = \frac{2\pi}{h^3c^2} \int_{E_{\min}}^{E_{\max}} \frac{E^2}{e^{(E-\Delta\mu)/k_BT} - 1} dE \tag{10.13}$$

representing the maximum absorbed or emitted *photon* flux density integrated over an energy band $E_{\min}$ to $E_{\max}$, and the function

$$L(E_{\min}, E_{\max}, T, \Delta\mu) = \frac{2\pi}{h^3c^2} \int_{E_{\min}}^{E_{\max}} \frac{E^3}{e^{(E-\Delta\mu)/k_BT} - 1} dE \tag{10.14}$$

representing the emitted *energy* flux density. Then we have for the current density,

$$J(V) = \frac{I(V)}{A} = q\{X f_s N(E_g, \infty, T_s, 0) + (1 - X f_s)N(E_g, \infty, T_a, 0)$$

$$- N(E_g, \infty, T_a, qV)\} . \tag{10.15}$$

The output *power* density, given by

$$P(V) = V \times J(V) \tag{10.16}$$

is now a function of E_g and V. For each value of E_g there exists some value of V, V_m, in the range 0 to E_g, for which $P(V)$ is a maximum, *i.e.*,

$$P_{\max} = P(V_m) \tag{10.17}$$

where

$$\left.\frac{\partial P}{\partial V}\right|_{V=V_m} = 0. \tag{10.18}$$

The *maximum* efficiency is then a function only of E_g and X and is given by

$$\eta_{max} = \frac{P(V_m)}{P_s} \tag{10.19}$$

where

$$P_s = X f_s L(0, \infty, T_s, 0) \tag{10.20}$$

represents the power received from the (black body) sun. For unconcentrated sunlight $(X = 1)$ η_{max} has a maximum of 31% at $E_g = 1.3$ eV [Araujo, 1994].

In Chapter 9 we saw how concentrating the light improves the balance between absorbed and emitted flux. At the maximum concentration, $X = 1/f_s$, and Eq. 10.15 becomes

$$J(V) = q\{N(E_g, \infty, T_s, 0) - N(E_g, \infty, T_a, qV)\} \tag{10.21}$$

leading to a maximum efficiency of around 41% at $E_g = 1.1$ eV. Calculated limiting efficiency as a function of band gap for a black body sun is shown in Fig. 10.4. The same analysis can be done using the standard AM1.5 solar spectrum, as shown in Fig. 9.10, and leads to very similar values for the limiting efficiency and optimum band gap.

It is straightforward to show that the same limit results if the angular range of emission is limited to the angular range of the sun. This may be achieved by inserting the cell inside a reflective cavity which admits solar and thermal radiation only through a single aperture, as shown in Fig. 10.3(a). In most of the examples considered in this chapter, we will consider only the case of full concentration, for clarity.

Summarising the assumptions made in the detailed balance analysis:

(i) All photons of energy greater than E_g, are absorbed to create one electron-hole pair. All photons with $E < E_g$ are not absorbed.
(ii) The electron and hole populations relax to form separate, distributions in quasi thermal equilibrium with the lattice at temperature T_a and with quasi Fermi levels separated by $\Delta\mu$.

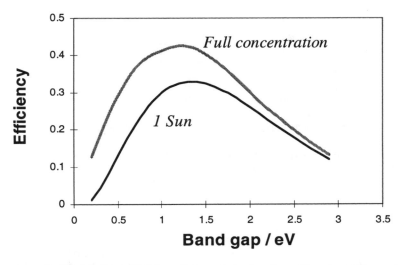

Fig. 10.4. Detailed balance limiting efficiency as a function of band gap, for a single band gap solar cell in a 5760K black body spectrum, at full concentration and at one-sun concentration levels. The maximum is similar to that obtained with the AM1.5 spectrum (Fig. 2.8).

(iii) Each electron is extracted with chemical potential energy μ such that $qV = \Delta\mu$. *I.e.*, the quasi Fermi level separation is constant throughout the device and equal to qV, where V is the potential difference at the terminals. This requires that carriers have infinite mobility.

(iv) The only loss process is spontaneous emission (also known as radiative recombination), which is required by detailed balance.

10.4. Multiple Band Gaps

The amount of work done per photon could clearly be increased if photons of different energies could be absorbed preferentially in cells of different band gap. A single band gap photoconverter functions most efficiently with monochromatic light which is tuned to the band gap. Then all photons are absorbed, and because the photon energy is close to the band gap, almost no electron kinetic energy is lost and the electrochemical potential extracted is close to the photon energy. If the solar spectrum could be split up and channelled into photoconverters of different band gaps, then more of the solar spectrum could be harnessed, each electron could be extracted with a chemical potential closer to the original photon, and a higher power could

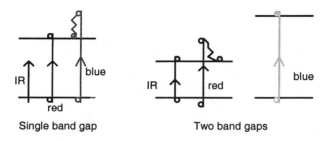

Fig. 10.5. Photon absorption in single gap and two gap system. The two gap system is capable of absorbing lower energy photons which are lost in the single band gap, and of extracting carriers generated by higher energy photons at an energy closer to the original photon energy.

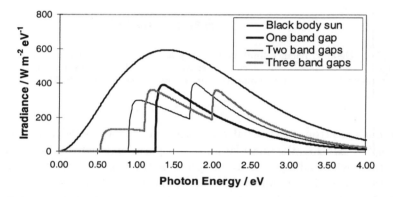

Fig. 10.6. Power available from optimised one, two and three band gap systems. (The different cells are assumed to be independently optimised.)

be extracted from the same spectrum. Figures 10.5 and 10.6 illustrate the point. Figure 10.7 illustrates one possible scheme for exploiting multiple band gaps, where sunlight is split up by means of dichroic mirrors and directed on to cells of different band gap.

10.5. Tandem Cells

10.5.1. *Principles of tandem cells*

In practice, efficient spectral splitting, as shown in Fig. 10.7, is hard to achieve. A more practical strategy is to stack different band gap junctions in optical series, and allow the wider band gap materials at the top to filter

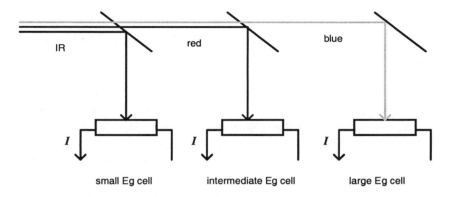

Fig. 10.7. Spectral splitting with dichroic mirrors allows different wavelength ranges to be absorbed preferentially in cells of different band gap.

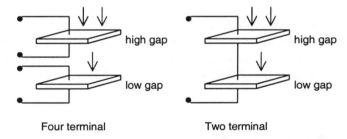

Fig. 10.8. Two and four terminal configurations for tandem cells. In either case, short wavelength light is preferentially absorbed in the top cell, and longer wavelength light in the bottom cell.

out most of the high energy photons, while less energetic photons pass through to smaller band gap materials below (Fig. 10.8). Greatest power is extracted if the output from the different junctions can be independently optimised. In the case of two band gaps, this is called a 'four terminal' tandem. The four terminal arrangement requires independent electrical contacts to top and bottom cell which is hard to achieve in practice. A more elegant arrangement is to connect the cells directly in series, so that a single current passes and voltages from the two cells are added. This 'two terminal' arrangement requires that currents from each cell be matched and constrains the performance so that the maximum output power is slightly less. Moreover, since current matching cannot be satisfied under all illumination conditions, the design is subject to additional, practical losses.

10.5.2. *Analysis*

We now calculate the power available using two band gaps, E_{g1} and E_{g2}, in a four terminal and a two terminal configuration.

In the four terminal configuration, the maximum output power is the sum of the maximum output powers from the two independent junctions. We assume that the spectral splitting is perfect, so that all photons with $E > E_{g2}$ are absorbed in the top cell, and all with $E_{g1} < E < E_{g2}$ in the bottom cell, where E_{g1} and E_{g2} are the band gaps of the bottom and top cell, respectively. Then the output power under maximum concentration is

$$P_{\max} = qV_{m1}\{N(E_{g1}, E_{g2}, T_s, 0) - N(E_{g1}, E_{g2}, T_a, qV_{m1})\}$$
$$+ qV_{m2}\{N(E_{g2}, \infty, T_s, 0) - N(E_{g2}, \infty, T_a, qV_{m2})\} \quad (10.22)$$

where V_{m2} is the voltage at which the top cell, and V_{m1} the voltage at which the bottom cell, produces maximum power. Provided that the two biases can be independently optimised, $P_{\max}$ is a function only of the two band gaps. Under full concentration, a maximum efficiency of over 55% is reached with band gaps of 0.75 and 1.65 eV, as shown in Fig. 10.9.

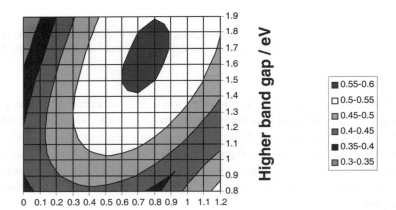

Lower band gap / eV

Fig. 10.9.   Calculated efficiency of a four terminal tandem under a 5760K black body sun in full concentration, as a function of the band gaps. A maximum of around 56% occurs in the central shaded region.

In the two terminal configuration the maximum power is given by the maximum of

$$P_{max} = q(V_1 + V_2)\{N(E_{g1}, E_{g2}, T_s, 0) - N(E_{g1}, E_{g2}, T_a, qV_1)\} \quad (10.23)$$

where V_1 and V_2 no longer represent the optimum voltages for the individual cells, but they satisfy the constraint of equal current

$$\{N(E_{g1}, E_{g2}, T_s, 0) - N(E_{g1}, E_{g2}, T_a, qV_1)\}$$
$$= \{N(E_{g2}, \infty, T_s, 0) - N(E_{g2}, \infty, T_a, qV_2)\}, \quad (10.24)$$

as well as the condition that P_{max} is stationary. The limiting efficiency in this case is slightly smaller, and more sensitive to the band gaps.

For either configuration, increasing the number of band gaps increases the limiting efficiency. In the limit of an infinite number of band gaps, the smallest band gap falls to zero and the limiting efficiency under one sun reaches 69%. Under full concentration, the limit approaches the thermodynamic limits of 86%.

10.5.3. *Practical tandem systems*

Tandem cells have been developed in a variety of materials combinations. Two terminal designs have been more widely studied than four terminal designs, because of the technologically appealing possibility of integrating different junctions in a single multilayer device by using 'tunnel junctions' to connect different p–n junctions. The tunnel junction is a heavily doped n–p junction which is generally assumed to introduce an Ohmic contact between the p terminal of one cell and the n terminal of the next. III–V materials are preferred on account of the high absorption coefficient and the possibility of tuning the band gap by compositional variation of ternary and higher alloys. In practice, the well understood binary semiconductor gallium arsenide (GaAs), is chosen for one of the cells in most cases. Although its band gap at $E_g = 1.42$ eV is different from the ideal for the top or bottom cell in a two junction tandem, the good material quality and carrier transport properties lead to better performance than poorer quality ternary materials with more suitable band gaps. Another important consideration is the compatibility of the different semiconductor materials in terms of their lattice constant, and thermal expansivity. Mismatched lattice constants lead to defective interfaces between the cells and enhance

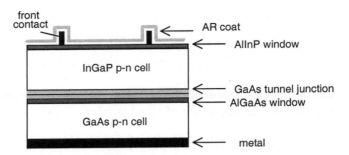

Fig. 10.10. Layer structure of an indium gallium phosphide/gallium arsenide two terminal tandem cell.

loss through recombination. Different coefficients of expansion lead to strain when the cell experiences changes in temperature.

The most efficient tandem cell produced to date is a 30.3% efficient indium gallium phosphide/gallium arsenide two terminal stack, developed by Japan Energy in 1996 (Fig. 10.10) [Green *et al.*, 2001]. Practical triple junction cells have been developed using the same two materials as the wider band gap components and the narrow band gap semiconductor, germanium, as the bottom cell. These have not yet surpassed the efficiency of the best double junction tandem. Tandem cells are expensive to manufacture, and are mainly being developed for use in space where efficiency is a premium.

10.6. Intermediate Band and Multiple Band Cells

10.6.1. *Principles of intermediate and multiple band cells*

The advantage of the tandem cell lies in the fact that the quasi Fermi level separation of the photogenerated electron and hole populations is closer, through the use of multiple band gaps, to the chemical potential of the exciting photon. A system which supports different quasi Fermi level separations delivers more work per incident photon. It would clearly be attractive if this could be achieved, not in several junctions made from different materials, but in a *single* junction using a *single* material. In a single band gap photoconverter, this is not possible because all of the conduction band states are coupled through phonon interactions and consequently all photogenerated electrons relax into a thermodynamic equilibrium with a single electrochemical potential, μ_n. All holes likewise relax into quasi thermal equilibrium with a single electrochemical potential, μ_p. Although it is possible to design a junction of variable band gap, preferentially absorbing photons of different energy at different spatial locations, phonon coupling of the spatially continuous band of electron states brings all photogenerated electrons into quasi thermal equilibrium with a single μ, and likewise for holes. In, the limit of an ideal, radiatively dominated, solar cell, the difference in μ's is dominated by the smallest band gap in the device [Araujo, 1994].

A hypothetical solution is to use a material which contains more than two bands. Figure 10.11 illustrates the case of an intermediate band system, containing two separate conduction bands. Under illumination, electrons are excited from the valence band into both the upper and intermediate conduction bands, and from the intermediate into the upper conduction band. Provided that the intermediate and conduction bands are not thermally coupled, then the electron populations in the different bands each form a local quasi thermal equilibrium with its own quasi Fermi level, $\mu_{n,i}$. The intermediate band may be introduced via impurities or quantum heterostructures which introduce electronic levels into the band gap, or it may be a result of the band structure.

Intermediate band solar cells have been proposed as hypothetical devices by several authors, [Kettemann, 1995; Luque, 1997; Green, 2000, 2001]. The idea of exploiting radiative transitions between intermediate levels is also a central concept of practical 'quantum well' solar cells [Barnham, 1991, 1997].

10.6.2. *Conditions*

We first explain the conditions for such a device to work, and then calculate the limiting efficiency.

(i) A condition for carriers to achieve independent quasi thermal equilibrium in a band is that collisions or scattering events *within* the band should be much more frequent than events *between* bands (discussed in Sec. 3.6). *This requires that there be a gap in the band structure which is large compared to the maximum phonon energy.* Otherwise, for a density of states which is continuous in energy, carriers can always be scattered into lower energy states by means of collisions with phonons.

(ii) *The feature which gives rise to the intermediate level should be periodic in space.* Periodicity imparts definite momentum to carrier states permitting spatial delocalisation, which aids transport. An imperfect array of isoenergetic impurities will lead to carrier localisation and poor transport. Moreover, the arrangement of levels into a band helps to suppress phonon scattering. For an intermediate band which lies within k_BT of the conduction or valence band, inter band scattering events are energetically allowed but are restricted by the requirement for momentum conservation. The restriction is lifted for an *isolated* impurity level because all momentum states are represented in the stationary wavefunction of the defect. Even in the case of a deep level which lies beyond k_BT from the nearest band edge, it is advantageous to form a band. Although single phonon scattering events to the ground state of the impurity are forbidden, in certain conditions *multiple* phonon scattering events may be allowed, whereby an electron is trapped by the impurity and then relaxes to the ground state by a series of phonon emissions, as the electron environment is successively altered by the presence of the electron. This successive relaxation and distortion would be symmetry forbidden in a periodic structure, and so multiple phonon emissions would not provide a route to the trapping of an electron by a band of deep levels.

(iii) For the intermediate band to be thermally isolated from the conduction band, it is necessary that electrons are extracted from only one of the bands. A selective contact should be made to the conduction band and not to the intermediate band. Otherwise the electron populations would be brought into thermal equilibrium through the contact. With this satisfied, the intermediate band is coupled to the valence and conduction bands only through optical transitions.

Analysis of the intermediate band cell

Let's assume that all the conditions for a system supporting separate quasi Fermi levels are satisfied, and calculate the limiting efficiency for the three band cell illustrated in Fig. 10.11. We will assume planar geometry and full concentration. There are three types of photon absorption event: photons of $E > E_{g2}$ which promote an electron from the valence band to **C1**; photons of $E > E_g$ which promote an electron from the valence band to **C2** and those of $E > E_{g1}$ which are absorbed to promote an electron from band **C1** to **C2**. Note that because **C1** and **C2** are both conduction bands and normally empty, the threshold for (strong) absorption is from the *bottom* of **C1** to the *bottom* of **C2**, and therefore $E_{g2} = E_g - E_{g1}$.) For the maximum amount of work per photon, a photon should only be absorbed across the

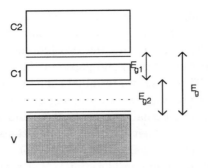

Fig. 10.11. Energy band diagram of an intermediate band material with a single valence band and two conduction bands, C1 and C2.

highest band gap which it is capable of crossing. This is the case if the absorption coefficient for the transition **V–C1** is much greater than for **C1–C2**, and that for **V–C2** much greater than for **V–C1**. Then each of the three types of absorption event occurs over a limited energy range: E_{g1} to E_{g2}; E_{g2} to E_g; and E_g to ∞. Assuming, as before, unit absorbance in each energy range, the output current density becomes

$$J(V) = q\{N(E_g, \infty, T_s, 0) - N(E_g, \infty, T_a, \mu_{c2} - \mu_v)\}$$
$$+ q\{N(E_{g2}, E_g, T_s, 0) - N(E_{g2}, E_g, T_a, \mu_{c1} - \mu_v)\} \quad (10.25)$$

where μ_v, μ_{c1} and μ_{c2} are the quasi Fermi levels of the valence, intermediate and conduction bands, respectively. In the steady state the net current from **V** to **C1** must be matched by the current from **C1** to **C2**, which yields the constraint,

$$q\{N(E_{g1}, E_{g2}, T_s, 0) - N(E_{g1}, E_{g2}, T_a, \mu_{c2} - \mu_{c1})\}$$
$$= q\{N(E_{g2}, E_g, T_s, 0) - N(E_{g2}, E_g, T_a, \mu_{c1} - \mu_v)\} \quad (10.26)$$

and the quasi Fermi levels must satisfy

$$(\mu_{c1} - \mu_v) + (\mu_{c2} - \mu_{c1}) = \mu_{c2} - \mu_v = qV . \quad (10.27)$$

For any combination of E_g and E_{g1}, there is some value of V which makes the power density, $V \times J(V)$, a maximum, subject to the above constraints. Optimum efficiency, the ratio of maximum output power density to input solar power density, is then a function only of the band gaps. For a 6000K

black body sun at full concentration, this efficiency has a maximum of
63.1% when $E_g = 1.93$ eV and $E_{g1} = 0.7$ eV [Luque, 1997].

This result can be compared to the two junction tandem cells where
η_{max} is likewise a function of two band gaps but the limiting value is much
smaller — around 55% for the two terminal configuration under maximum
concentration. There are different ways to regard the comparison. One is
that the multi band cell actually presents *three* and not two band gaps
to the incoming photons, albeit with some limitations on the band gaps
and the Fermi level separations, and should be compared with the three
junction tandem cell. Another consideration is that with a series connected
tandem cell, two photon absorption events are always required to deliver
one electron to the external circuit, while with the intermediate band cell
either two or one photons are needed, depending on photon energy, so the
photocurrent quantum efficiency is higher. In any case, the intermediate
band cell is based upon a technically more challenging materials situation,
on account of the requirements for a thermally isolated intermediate band
and selective contacts.

10.6.3. *Practical strategies*

A variety of practical routes to the intermediate band solar cell have been
proposed. The simplest is the idea that intra band states, introduced by
ambivalent impurities in the semiconductor, can take part in optical transitions, and absorb photons of lower energy than the band gap to accept an
electron from the valence band (if empty) or to deliver an electron to the
conduction band (if full), as shown in Fig. 10.12. Enhanced photocurrent
at sub band gap energies has in fact been observed, in silicon doped with
indium [Keevers, 1994]. However, it has not been shown, and it is unlikely,
that such a system could support independent quasi Fermi levels, since the
isolated impurities may act as centres for non radiative recombination via
multiple phonon events.

Another proposal is to use materials which naturally possess multiple
bands of narrow width. Examples are polar semiconductors, such as II–VI
materials or semiconducting oxides, and molecular semiconductors, which
have conduction band widths typically of some tenths of eV. Narrow band-
width materials are also attractive for tandem devices, on account of their
finite band widths, which enhance the photon selectivity in different materials. Finite band widths are offered by dye molecules, which can be imple-
mented as sensitisers on non absorbing conducting substrates.

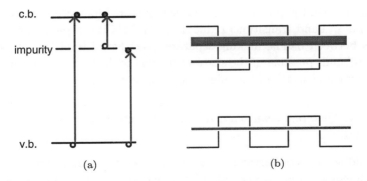

Fig. 10.12. (a) Impurity band semiconductor; (b) Quantum dot minibands.

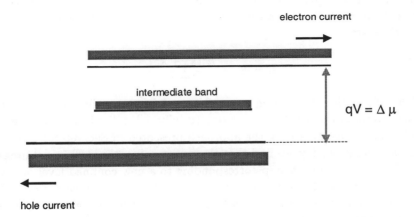

Fig. 10.13. Schematic energy-space diagram of intermediate band cell. Selective contacts are needed so that the intermediate band can be isolated from the external circuit, and the extracted voltage determined by the splitting of conduction and valence band Fermi levels.

A more elegant but ambitious proposal is to engineer materials with the desired band structure using low dimensional semiconductor heterostructures. Complete discretisation of the electronic energy levels can be achieved using quantum dots. In these nanometre-sized semiconductor particles the carriers are confined to three dimensions, resulting in an atomic like band structure of discrete energy levels. If the quantum dot is grown within a second, wider band gap semiconductor material, then the dot introduces a discrete level within the band gap of the host semiconductor. A *regular* array or *superlattice* of identical quantum dots leads to a set of minibands within the band gap of the host semiconductor. Momentum conservation for band — band transitions reduces the probability of phonon scattering events between bands, making separate quasi Fermi level for electrons in the minibands possible. If the size and spacing of the dots can be controlled, the energy and width of the minibands can be tuned.

Box 10.1. Quantum semiconductor heterostructures for multiple band photovoltaics

Building semiconductor heterostructures on the nanometre scale leads to confinement of electrons and holes in one, two or three dimensions, and leads to densities of states which reflect the symmetry of the confining structure. For a quasi two-dimensional system, such as a 'quantum well', the density of states is quantised in one direction, so that the electron has kinetic energy

$$E = E_n + \frac{\hbar^2(k_x^2 + k_y^2)}{2m^*}$$

where E_n is the energy due to confinement in the z direction, and the second term is the kinetic energy due to fully delocalised motion in the x and y directions. Following the approach presented in Sec. 3.3 for a three-dimensional geometry, we find that the density of states $g(E)$ is a staircase like function, with each step corresponding to a new confined level or sub-band E_n. The positions of these steps can be tuned by the choice of 'well' and 'barrier' materials and the width of the well, allowing engineering of the band structure. When carriers are confined in three directions, a quantum dot results, with discrete set of energies like a super atom. A regular array of such dots allows the discrete QD spectra to merge into bands. The dot dimensions determine the positions of the bands and the array spacing determines the band width, so that the band structure can be completely designed. Such systems are the basis of ideas for multiple band solar cells.

Quantum well structures are less suitable for intermediate band structures because the QW density of states is continuous in energy. A carrier in one sub-band can always be scattered into a lower band, (with different in-plane momentum) by phonon interaction. Nevertheless, QWs have already been exploited for multiple band photovoltaic devices by embedding a set of quantum wells within the depletion region of a *p–i–n* structure. The QWs introduce a new band gap so that photons of low energy may be absorbed, increasing the current, while still allowing high energy photons to be efficiently absorbed in the host material. QW devices appear to benefit from a difference in quasi Fermi level separation between the wide band gap and low band gap regions, in contrast to the assumptions of detailed balance theory that $qV = \Delta\mu$ at all points. The reasons for this and the implications have yet to be clarified.

In addition, the loss of symmetry in electron and phonon density of states in QWs may make quasi 2D systems useful for hot carrier solar cells, discussed below.

10.7. Increasing the Work Per Photon using 'Hot' Carriers

Multiple band gap approaches are based on capturing photons of different energy in materials of different band gap, and extracting the photogenerated carriers with a chemical potential related to the band gap of the absorbing material used. An alternative approach is to increase the work done per photon by harnessing some of the excess kinetic energy of the photogenerated carriers before they relax. This could be done if electron phonon interactions could be slowed down so that the photogenerated carriers can be collected while still 'hot', or if the excess kinetic energy of hot carriers can be exploited to generate more carrier pairs by a process known as *impact ionisation*. The first results in an increased voltage and the second in an increased photocurrent. Both rely on similar physics and lead to identical limiting efficiencies, but we will treat them separately here, since the routes have been proposed on account of different physical observations.

10.7.1. *Principles of cooling and 'hot' carriers*

Let's first consider the sequence of events following photon absorption. At the instant of excitation, the photogenerated carriers have a kinetic energy distribution which reflects the distribution of photon energies from the sun and the absorption spectrum of the material. Elastic carrier–carrier scattering ('equilibriation') over the next few hundred femtoseconds brings each carrier population into a self equilibrium which can be described by a temperature, T_H, and a chemical potential, μ_H. T_H is greater than the ambient temperature and in certain cases can be equal to T_s.

Over the next few picoseconds, carriers in the hot plasmas lose kinetic energy through collisions with phonons until the temperature of the distribution is reduced to the lattice temperature, T_a ('cooling'). Then each carrier distribution is in quasi thermal equilibrium with the lattice, and

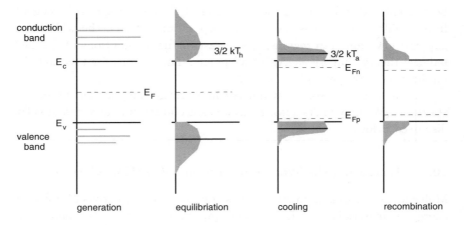

Fig. 10.14.   Evolution of the carrier energy distributions following photogeneration.

the phonon absorption rate is balanced by phonon emission. However, the electron and hole populations are not yet in thermal equilibrium with each other. The populations of electrons in the conduction band, and of holes in the valence band, are increased over the equilibrium populations in the semiconductor in the dark, as though the electron Fermi level had moved towards the conduction band, and the hole Fermi level towards the valence band, as shown in Fig. 10.14. This Fermi level *splitting* is equal to the chemical potential of the radiation, and increases with light intensity. During cooling, the carriers give up kinetic energy as heat to the lattice, and the entropy of the system increases.

Over a longer time scale the carriers *recombine* by radiative and non radiative routes. In the current context we ignore nonradiative processes which are assumed to be avoidable. (Auger processes are considered in the next section.) Radiative recombination occurs over nanoseconds to microseconds, depending on the absorption coefficient of the material. The effect of recombination is to reduce the electron and hole populations and move the quasi Fermi levels back towards the equilibrium position. Under constant illumination, populations achieve a steady state where the photogeneration rate is equal to the sum of the recombination rate and the rate of carrier extraction (as electric current). This balance determines the electrochemical potential and the power output of the cell. For efficient energy conversion the rate of carrier extraction should be comparable with or faster than the rate of recombination. In the conventional picture, the first two processes of equilibration and cooling are assumed to be complete for any carrier before it is extracted, and only recombination needs be considered in determining the power output of the cell.

The central concept of the hot carrier solar cell is that cooling *need not* be complete before carriers are extracted. If extraction can be speeded up, or cooling slowed down, then some of the carriers can be extracted while still hot and deliver some of their excess kinetic energy to the electrical output of the cell. These conditions could arise in systems of unusual electronic and phononic structure. (Cooling is influenced by the availability of both phonons of appropriate energy and momentum to absorb electronic kinetic energy and of lower energy states for the carriers to be scattered into.) A further requirement is that the contacts are designed to allow adiabatic cooling of hot carriers during extraction. Some proposed practical schemes are discussed below.

10.7.2. *Analysis of the hot carrier solar cell*

To calculate the maximum efficiency of the hot carrier solar cell we consider the limit where there is *no* carrier–lattice scattering. We consider a material of band gap E_g and make the usual assumptions for the limiting photovoltaic efficiency: that all photons of $E > E_g$ are absorbed; that each photon generates one electron-hole pair; and that radiative recombination is the only loss process. As usual, illumination is provided by a black body sun at temperature T_s and geometric factor F_s. (The formulation of this section is based on [Wuerfel, 1997] and [Ross, 1982].)

After photogeneration, each carrier population equilibrates through carrier–carrier scattering. At equilibrium, the Gibbs free energy is minimised and therefore the quantity

$$\sum_i \eta_i dn_i$$

is conserved through any scattering event, where η_i is the electrochemical potential of carrier i and dn_i the particle number. For the electron scattering event $e_1 + e_2 \rightarrow e_3 + e_4$, it follows that

$$\eta_{e1} + \eta_{e2} = \eta_{e3} + \eta_{e4} \tag{10.28}$$

since $dn_i = 1$ for each carrier. Since carrier scattering is elastic, the kinetic energies obey

$$E_{e1} + E_{e2} = E_{e3} + E_{e4} \tag{10.29}$$

These equations are satisfied for all carrier pairs when η_i depends on kinetic energy through

$$\eta_i = \eta_0 + \gamma E_i \tag{10.30}$$

where γ and η_0 are constants. If $\gamma = 0$ we have the familiar case that all carriers at equilibrium have the same electrochemical potential. In the absence of electron-lattice scattering, the solution $\gamma > 0$ is also valid [Wuerfel, 1997]. Assuming a similar relation for the holes (in principle the value of γ may be different for holes), we have for the chemical potential of the electron-hole pair

$$\Delta\mu = \eta_{e0} + \eta_{h0} + \gamma(E_e + E_h) \tag{10.31}$$

Hence the chemical potential of an electron-hole pair generated by a photon of energy E is given by

$$\Delta\mu = \mu_0 + \gamma E \tag{10.32}$$

where

$$\mu_0 = \eta_{e0} + \eta_{h0} - \gamma E_g . \tag{10.33}$$

The constants η_{e0}, η_{h0} depend upon the extraction conditions.

Now, by considering interaction with photons of energy E, the carriers will be distributed according to the function

$$f = \frac{1}{e^{(E-\Delta\mu)/k_B T_a} + 1} \tag{10.34}$$

where T_a is the (ambient) temperature and $\Delta\mu$ the energy dependent chemical potential, Eq. 10.32 above. By rewriting f as

$$f = \frac{1}{e^{E(1-\gamma)/k_B T_a - \mu_0/k_B T_a} + 1} \tag{10.35}$$

it is evident that the same distribution can be described by a *single* chemical potential, μ_H, if the temperature is defined as

$$T_H = \frac{T_a}{1 - \gamma} . \tag{10.36}$$

Then

$$\Delta\mu = \mu_H \frac{T_a}{T_H} + E\left(1 - \frac{T_a}{T_H}\right). \qquad (10.37)$$

(The same relations follow by considering the entropy change per absorbed photon.) This means that the luminescence from the carrier distribution is exactly equivalent to that from a distribution at temperature T_H and chemical potential μ_H, and so it is meaningful to consider the carriers as being at a hot temperature T_H.

This enables us to calculate the current output. As usual in the radiative limit, J is given by the difference between the net absorbed and emitted photon flux

$$J(V) = q\{Xf_s N(E_g, \infty, T_s, 0) - N(E_g, \infty, T_H, \mu_H)$$

$$+ (1 - Xf_s)N(E_g, \infty, T_a, 0)\}. \qquad (10.38)$$

Note that in the case of full concentration, Xf_s becomes equal to 1 and the final term in Eq. 10.38 vanishes.

For the hot carrier device, the dependence of V on μ_H is no longer trivial. On extraction, the carriers must cool to the ambient temperature T_a and in so doing the chemical potential of the extracted carriers will increase. For maximum work, this cooling should be done at constant entropy. This is possible if electrons and holes are withdrawn through separate contacts of extremely narrow energy range, $\Delta E \ll k_B T$, so that no kinetic energy is lost during cooling. At constant entropy Eq. 10.37 is valid, and the chemical potential of the extracted carriers is given by

$$\mu_{out} = \mu_H \frac{T_a}{T_H} + E_{out}\left(1 - \frac{T_a}{T_H}\right) \qquad (10.39)$$

where E_{out} is the energy separation of the electron- and hole-selective contacts. Thus the voltage output is $V = \mu_{out}/q$ and the extracted power is

$$P(V) = \frac{\mu_{out}}{q} \times J. \qquad (10.40)$$

This description is valid so long as carrier extraction is slow compared to carrier–carrier scattering. It is assumed that the equilibrium distribution of carriers is unaffected by the extraction of carrier pairs. Fast carrier scattering replenishes the population of carrier pairs with energy E_{out} as the carriers are extracted.

One more condition is required to calculate the power output. This comes from the conservation of energy within the hot device. Since energy is introduced only through photon absorption and is lost only through power extraction and hot carrier emission, we can write

$$JE_{out} = q\{Xf_sL(E_g, \infty, T_s, 0) - L(E_g, \infty, T_H, \mu_H)$$

$$+ (1 - Xf_s)L(E_g, \infty, T_a, 0)\}. \tag{10.41}$$

We can now calculate the current–voltage characteristics of the system. The performance of the hot carrier systems depends upon *three* externally determined, or design, parameters: E_{out} as well as the usual parameters of E_g and V. The value of E_{out} importantly influences the performance of the device as a solar cell.

For a given E_g and E_{out} the device operates as a solar cell over a range of V from 0 (short circuit) to V_{oc}, where $J = 0$. Varying the output voltage controls the temperature of the carriers. At short circuit where all photo-generated carriers are collected, $\mu_{out} = 0$, and $T = T_a$ in all cases; *i.e.*, there is no hot carrier effect. As V is increased towards V_{oc}, J decreases due to increased radiative recombination, and T_H increases towards a maximum value which is close to T_s at open circuit. During this cycle, the behaviour of μ_H is rather counterintuitive: μ_H becomes more negative as the output voltage is increased, unlike the usual case where $\Delta\mu$ increases exactly as V.

It may be helpful to think of this behaviour in terms of energy balance, since the photogenerated carriers introduce excess kinetic energy which is conserved during elastic scattering and not lost by cooling. At short circuit, all carriers introduced by the light are extracted, and so contribute no excess kinetic energy to the carrier distribution in the solar cell. Therefore $T_H = T_a$. As output voltage is increased, the carrier density in the cell increases, the kinetic energy per carrier increases on account of the more energetic photogenerated carriers, and so the temperature of the carrier distribution increases. At open circuit where no carriers are extracted, the photogenerated carriers contribute the maximum kinetic energy and T_H has its maximum value. In the limit of full concentration and small band gap, T_H at open circuit is equal to T_s.

Once E_g and E_{out} are determined, the current–voltage characteristic for the hot carrier device can be found by varying V between 0 and V_{oc}. The maximum power point will correspond to certain values of T_H and μ_H. A family of J–V curves are illustrated in Fig. 10.15. Notice that the curve is

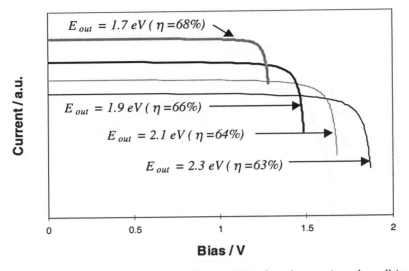

Fig. 10.15. Calculated current–voltage characteristics for a hot carrier solar cell in an unconcentrated black body sun. Curves for different values of the extraction energy, E_{out}, are shown, and the maximum efficiencies for each curve are given. Notice how the knee of the curve is almost square in the highest efficiency case.

much squarer than those for the radiative limit, consistent with a higher efficiency. In the radiative limit, the curvature of the J–V curve is due to the radiative current rising like $\exp(V/k_BT)$ at constant T. For the hot carrier cell, although losses are still radiative, the temperature is increasing as V increases, and the shape no longer described by a single T.

Efficiency can be calculated as a function of the design parameters, E_g and E_{out}. For comparison with the radiative limit, it is convenient to make E_{out} a variable parameter and find the values of V and E_{out} which lead to the highest output power for each E_g. Those values will correspond to particular values of T_H and μ_H. (Indeed, the calculation simplifies if we consider T_H and μ_H as optimisable parameters rather than the design and operation variables of E_{out} and V.) Figure 10.16 shows the maximum efficiency as a function of E_g, together with the corresponding values of E_{out}, T_H, and μ_H. For full concentration, η has a maximum of $\sim 85\%$ at a temperature of 3600K and a band gap of a few tenths of eV. For an unconcentrated black body sun η has a maximum of $\sim 65\%$. The temperature for maximum power is unaffected by concentration, in contrast to the case of impact ionisation, discussed below.

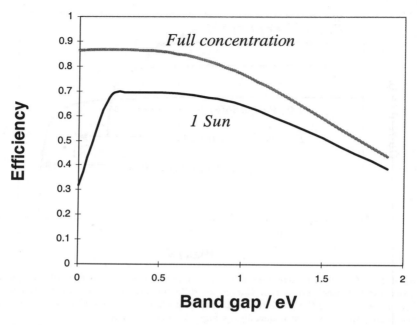

Fig. 10.16.   Calculated efficiency as a function of band gap for hot carrier solar cell
under full concentration and 1-sun illumination by a 5760K black body sun.

10.7.3. *Practical strategies*

There are two materials requirements for hot carrier devices: a photoactive
material where cooling is slower than transport to the contacts; and contact materials which permit selective electron or hole extraction through a
narrow energy band.

   In typical bulk semiconductor devices, cooling occurs in less than 10
ps, while carrier extraction may take nanoseconds or longer. The rate of
cooling depends upon the availability of phonons of appropriate energy and
momentum to scatter the excited carriers into lower energy states. Phonons
fall into two groups: acoustic phonons, which typically have energy of a few
meV at room temperature and longitudinal optical (LO) phonons which
have a higher energy — typically 30–40 meV in III–V semiconductors at
room temperature. LO phonons are most likely to be involved in cooling hot
carriers. The phonon momentum distribution in a bulk material is approximately isotropic but related to the crystal band structure. Since the density
of electron states in a bulk semiconductor is continuous, and isotropic, a

carrier in an excited state has a high probability of being scattered into an unoccupied, lower energy state by a suitable phonon. This can be prevented if (i) there is a shortage of phonons or (ii) there are no available electron states within a phonon energy. The first situation arises under very intense illumination. At high enough injection levels, there may not be enough phonons for all the photogenerated carriers to relax promptly, and in the steady state the energy distribution of the carriers is hot. In addition, the strong interactions of higher energy phonons with the hot carrier distribution disturbs the equilibrium of the phonon distribution, so that the phonons themselves may become hot, and less effective at cooling. However, this requires very high light intensities, orders of magnitude more intense than the sun.

Slowed cooling may be observed more easily in semiconductor structures of reduced dimension. In quantum wells, carriers are confined into two dimensional slabs and the electronic structure is quantised into sub-bands. Although the density of states is still continuous, it is highly anisotropic and an excited carrier may only be scattered into a lower energy level by a phonon of the correct momentum (for intra sub-band transitions, the phonon wavevector must lie in the plane of the QW). Greatly increased cooling times and hot electron spectra have been observed in QWs of GaAs in $Al_xGa_{1-x}As$, with cooling times of up to 1000 ps compared to less than 10 ps for bulk GaAs, at light intensities of 10^4 suns. (The carrier temperature can be observed experimentally from the energy distribution of the photoluminescence spectrum.) For a review, see Nozik [Nozik, 2000].

Quantum *dots* (QDs) are potentially more useful. Carriers in a QD are confined in all dimensions and the electronic structure is quantised into discrete levels. If the spacing of these levels is larger than the energy of the most energetic phonon, then scattering of a carrier from an excited to a lower energy state through a single phonon interaction is impossible. Carriers could only relax via multi phonon interactions which are expected to be relatively very slow. This is the phonon bottleneck effect, mentioned above in the context of intermediate band solar cells. Slowed cooling has been observed in QD structures, again under high injection conditions, but not to a greater extent than in QWs, and the predicted phonon bottleneck is not seen. In practice, alternative relaxation mechanisms, such as Auger recombination and trap assisted relaxation, may operate.

An alternative strategy is to make use of structures where the charge separation is extremely fast. Interfacial charge transfer at semiconductor electrolyte junctions and in heterogeneous molecular systems can be very fast: for instance, in dye sensitised solar cells, electrons are injected from a molecular sensitiser into a bulk semiconductor in less than a picosecond [Hagfeldt, 2000]. The injected carriers are inevitably hot, and this could be exploited if they could be collected before cooling. This would require very thin devices, to reduce the transit time of the injected carrier to the contacts. The challenge will be to combine high optical absorption with short transit times.

The second requirement is selective contacts with a narrow energy band. Usual contacting materials, such as metals or heavily doped semiconductors, are not suitable but low dimensional semiconductor structures may again be useful. One possibility is to a superlattice structure, where the density of states is quantised into bands, which may be made narrow and well separated by choosing the materials and periodicity carefully.

In the analysis of *practical* hot carrier systems, cooling by phonon interactions cannot be ignored. The drift-diffusion and carrier continuity equations which are used to analyse normal semiconductor devices are based upon *thermalised* carrier populations and do not apply. To allow for spatial variations in the carrier temperature, a further continuity equation must be introduced. This is the energy balance equation. In the steady state the electron kinetic energy flux S_n obeys

$$\nabla S_n = J_n F + W_{\text{relax}} + W_{\text{gen}} + W_{\text{rec}} \qquad (10.42)$$

where the first term represents the gain in kinetic energy through electron acceleration by the electric field, W_{relax} is the loss of kinetic energy by relaxation, W_{gen} the loss by electron recombination and W_{gen} the generation of kinetic energy by optical generation. A similar equation applies for hot holes. These should be solved along with the current and continuity equations and Poisson's equation, leading to profiles for the carrier kinetic energy (which is effectively the carrier temperature, since kinetic energy $= 3/2k_B T$) as well as for carrier density, current density and electric field. Note that in defining the current, the contribution due to a gradient in carrier temperature (the Seebeck effect) must be included. For further details see [Lunsdtrom, 1990; Hanna, 1997.]

10.8.   Impact Ionisation Solar Cells

The final route, to be discussed here, to increasing the work done per photon is impact ionisation. This is a scheme where a relaxation process is introduced which competes with cooling and leads to further carrier pair generation. In the following analysis we will see that the impact ionisation solar cell is closely related to the hot carrier solar cell.

Impact ionisation, or Auger generation, is the reverse of Auger recombination. In Chapter 4 we saw that Auger recombination is a three body process, where an electron collides with a second electron, or with an impurity, recombines with an available hole and gives up its electrochemical potential energy as kinetic energy to the second electron. In the reverse process, an energetic electron collides with the lattice and gives up its kinetic energy to excite a further electron across the band gap. In the context of a photovoltaic device, this means that the quantum efficiency for light with $E > 2E_g$ can be greater than one. These high energy photons are capable of *multiple pair generation*. This effect has been observed, and attributed to Auger generation, in germanium photodiodes at photon energies $E > 2.5$ eV ($E_g = 0.7$ eV) and in silicon diodes for $E > 3.3$ eV ($E_g = 1.1$ eV) [Werner, 1995].

In Auger processes energy and momentum must be conserved. Suppose an electron e_1 with kinetic energy E_{e1} and momentum k_{e1} relative to the symmetry point k_0 causes generation of a pair e_2, h_2. Then

$$k_{e1} = k'_{e1} + k_{e2} - k_{h2} \qquad (10.43)$$

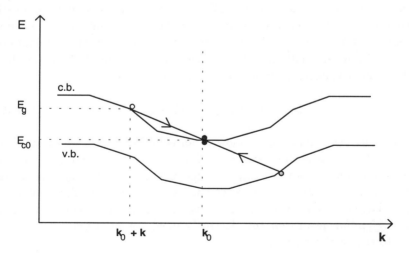

Fig. 10.17. Energy-momentum band structure diagram showing the conditions for Auger generation.

and

$$E_{e1} = E'_{e1} + E_{e2} - E_{h2}. \tag{10.44}$$

Auger generation events depend on availability of electron and hole states at correct $\mathbf{k}$ and E and therefore the band structure is most important in determining the probability of generation events. The conditions on the band structure for effective Auger generation have been discussed by Werner *et al.* [Werner, 1994] and are summarised at the end of this section.

10.8.1.   *Analysis of impact ionisation solar cell*

In the following paragraphs we develop the theory for the limiting efficiency of a cell admitting multiple pair generation, as developed by Wuerfel [Wuerfel, 1997]. The different analysis proposed by the inventors, Werner, Brendel, Kodolinski and Queisser, has subsequently been shown to be flawed, in a thermodynamic respect. This original analysis is appended at the end of this section.

To calculate the maximum efficiency for an impact ionisation solar cell, we will assume that all the conditions for multiple pair generation are satisfied. The reverse process of impact ionisation, Auger recombination, must be included to satisfy detailed balance. We will assume that carrier relaxation by phonon interactions is sufficiently slow that it can be ignored. Therefore, the photogenerated carriers are allowed to equilibriate, but not to cool, and they may recombine only by radiative or Auger recombination. We know from Eq. 10.31 that the electrochemical potential of an equilibriated carrier pair depends upon the carriers' energy as

$$\Delta\mu = \eta_{e0} + \eta_{h0} + \gamma(E_e + E_h). \tag{10.45}$$

When Auger recombination is allowed, an equilibrium will be established between Auger generation and recombination, where a number dn_1 of electron-hole pairs of energy E_{eh1} and chemical potential $\Delta\mu_{eh1}$ are transformed into a number dn_2 pairs of energy E_{eh2} and chemical potential $\Delta\mu_{eh2}$. The conditions of energy conservation and minimisation of the free energy $\sum_i \mu_i dn_i$ mean that the electrochemical potential must be *proportional* to the carrier energy, hence

$$\Delta\mu = \gamma(E_e + E_h).$$

Considering the luminescence from such a distribution we find, as before (Eq. 10.36), that the distribution can be described by a raised temperature

$$T_{\mathrm{H}} = \frac{T}{1 - \gamma}$$

and a chemical potential $\mu_{\mathrm{H}} = 0$. Zero chemical potential is expected for a thermal distribution where particle number is not conserved.

In calculating the power conversion efficiency we proceed as in the hot carrier case, except that the current can no longer be related to the difference of solar generation and radiative recombination (because additional current is generated through impact ionisation). Instead, the output current is calculated from the energy balance

$$JE_{\mathrm{out}} = q\{Xf_{\mathrm{s}}L(E_{\mathrm{g}}, \infty, T_{\mathrm{s}}, 0) - L(E_{\mathrm{g}}, \infty, T_{\mathrm{H}}, 0)$$

$$+ (1 - Xf_{\mathrm{s}})L(E_{\mathrm{g}}, \infty, T_{\mathrm{a}}, 0)\} \tag{10.46}$$

and the power density is given by

$$P = JV = q\left(1 - \frac{T_{\mathrm{a}}}{T_{\mathrm{e}}}\right)\{Xf_{\mathrm{s}}L(E_{\mathrm{g}}, \infty, T_{\mathrm{s}}, 0) - L(E_{\mathrm{g}}, \infty, T_{\mathrm{H}}, 0)$$

$$+ (1 - Xf_{\mathrm{s}})L(E_{\mathrm{g}}, \infty, T_{\mathrm{a}}, 0)\} \tag{10.47}$$

where the output voltage, V, is related to the extraction energy E_{out} through

$$V = \frac{1}{q}\mu_{\mathrm{out}} = E_{\mathrm{out}}\left(1 - \frac{T_{\mathrm{a}}}{T_{\mathrm{H}}}\right). \tag{10.48}$$

The energy balance, expressed by Eq. 10.47, is equivalent to that used in the hot carrier solar cell. It is still valid because Auger processes conserve energy, although they do not conserve particle number.

Although three design parameters (E_{g}, E_{out} and V) appear to be needed to find the J–V characteristic of an impact ionisation cell, only two of these are independent. Notice that P depends only on two parameters, E_{g} and T_{H}, the second of which can be controlled by varying either V or E_{out}. This means that for the impact ionisation solar cell, the extraction energy E_{out} makes no difference to the maximum efficiency. As with the hot carrier solar cell, T_{H} varies from its minimum value of T_{a} at short circuit, to its maximum value at open circuit as V is increased. The value of T_{H} at the maximum power point depends upon the level of illumination.

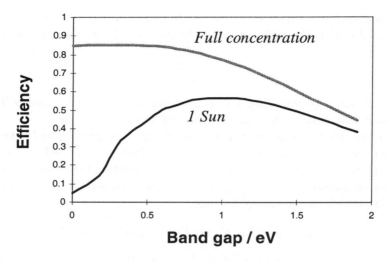

Fig. 10.18. Efficiency as a function of band gap for impact ionisation solar cell under 1-sun and full concentration.

Power conversion efficiency is plotted as a function of band gap in Fig. 10.18. For an unconcentrated black body sun, efficiency has a maximum of 55% at a band gap of around 1 eV. Under full concentration, the efficiency reaches a maximum of 85% at $T_H = 2460$ and a small, vanishing band gap. The behaviour under full concentration is identical to that of a fully optimised hot carrier cell. Once again, this maximum performance under full concentration corresponds to the thermodynamic limit.

It is interesting to note that, except in the case of full concentration, impact ionisation leads to a lower limiting efficiency than simply preventing cooling, as in the case of the hot carrier cell. This suggests that a band structure conducive to impact ionisation may not be a desirable property of a high efficicency photovoltaic material. However it has the advantage of easier design conditions (E_{out} does not need to be optimised to obtain the maximum performance). Moreover, in realistic systems, impact ionisation needs to be considered in competition with relaxation by phonon interactions. In such conditions, the hot carrier device would not necessarily outperform the impact ionisation device, even at less than full concentration.

An alternative analysis of the impact ionisation solar cell was proposed by the inventors [Werner, 1994, 1995]. In their theory, it is assumed that the electron and hole populations are in thermal equilibrium with the lattice, at temperature T_a. This assumption is incompatible with suppressed carrier cooling. This approach also arrives at a photon energy dependent chemical potential, but for a different reason, and leads to identical limiting efficiency of 85% under full concentration as $E_g \rightarrow 0$. The approach differs from the above in the models of recombination: above we consider that a *single* hot electron-hole pair with chemical potential $\mu = \mu E$ recombines to produce a photon of energy E. In the WKBQ model, *multiple* carrier pairs at $T = T_a$ (with $\Delta \mu = qV$) recombine to give a single photon. This multiple carrier recombination process is physically very unlikely.

Practical strategies

Since suppressed phonon interactions are a condition for impact ionisation solar cells, the practical strategies discussed above in the section on hot carrier devices apply, as does the requirement for contact materials with very narrow energy ranges, to allow isoentropic cooling.

In addition, the electronic band structure should favour impact ionisation. The conditions for this are discussed by Werner *et al.* [Werner, 1994]. They include the requirement for direct points in the band structure with band gap greater than twice the fundamental band gap; the requirement that valence band and conduction band be parallel over wide ranges of **k** (this allows carrier pairs to be generated so that only one carrier possesses the excess energy); and the requirement for strong light absorption at points which permit Auger generation. The first two of these imply that the material should have an indirect band gap, as do silicon and germanium. Finding the atomic or crystal configuration which generates the ideal band structure for multiple pair generation is an inverse problem, (inverse of solving the Schrödinger equation) and so far has not been attempted.

10.9. Summary

The limiting efficiency for a single junction solar cell, from considerations of the detailed balance between absorbed and emitted radiation, is around 31% in unconcentrated sunlight. In principle, this limit can be increased by relaxing one or more of the assumptions on which this calculation is based. In this chapter we have presented a number of routes to increased efficiency. These are based on: (1) the preferential absorption of photons of different energy in materials of different band gap, which is the basis of tandem and multiple band solar cells; (2) the exploitation of radiative transitions between the principal valence and conduction bands and an intermediate band; (3) the rapid collection of photogenerated carriers to make use of their kinetic energy before they reach thermal equilibrium with the environment, and (4) the generation of multiple carrier pairs by absorbed photons with energy greater than the band gap.

All of these strategies place exceptional demands on the materials used. Ideally, high quality materials are needed so that non-radiative recombination can be eliminated. Materials of extraordinary electronic and optical properties are needed to engineer the required band structure. Nanostructured materials, including low dimensional semiconductor structures and quantum dots, are cited as possible ways of realising the hypothetical structures.

References

G.L. Araujo and A. Marti, "Absolute limiting efficiencies for photovoltaic energy-conversion", *Sol. Energy Mater. Solar Cells* **33**, 213 (1994).

K.W.J. Barnham, B. Braun, J. Nelson and M. Paxman, "Short-circuit current and energy efficiency enhancement in a low-dimensional structure photovoltaic device", *Appl. Phys. Letts.* **59**, 135–137 (1991).

K. Barnham *et al.*, "Quantum well solar cells", *Appl. Surf. Sci.* **114**, 722–733 (1997)

A. de Vos, *Endoreversible Thermodynamics of Solar Energy Conversion* (Oxford University Press, 1990).

A. Hagfeldt and M. Graetzel, "Molecular photovoltaics", *Acc. Chem. Res.* **33**, 269–277 (2000), and references therein.

M.A. Green, "Prospects for photovoltaic efficiency enhancement using low-dimensional structures", *Nanotechnology* **11**, 401–405 (2000).

M.A. Green, "Multiple band and impurity photovoltaic solar cells: General theory and comparison to tandem cells", *Progr. Photovoltaics* **9**, 137–144 (2001).

M.A. Green, K. Emery, D.L. King, S. Igari and W. Warta, "Solar cell efficiency tables", *Progr. Photovoltaics* **9**, 287 (2001).

M.C. Hanna, Z. Lu and A.J. Nozik, "Hot carrier solar cells", *Proc. 1st NREL Conference on Future Generation Photovoltaic Technologies*, AIP Conference Proceedings 404 (1997).

M.J. Keevers and M.A. Green, "Limiting efficiencies of ideal single multiple energy gap terrestrial solar cells", *J. Appl. Phys.* **75**, 4022 (1994).

S. Kettemann and J.-F. Guillemoles, "Limiting efficiency of LDS solar cells", *Proc. 13th European Photovoltaic Solar Energy Conference*, 119 (H.S. Stephens and Associates, Bedford, 1995).

M. Lundstrom, *Fundamentals of Carrier Transport* (Wokingham: Addison-Wesley, 1990).

A. Luque and A. Marti, "Increasing the efficiency of ideal solar cells by photon induced transitions at intermediate levels", *Phys. Rev. Lett.* **78**, 5014 (1997).

A.J. Nozik, "Spectroscopy and hot electron relaxation dynamics in semiconductor. Quantum wells and quantum dots", *Ann. Rev. Phys. Chem.* (submitted) 2001.

R.T. Ross and A.J. Nozik, "Efficiency of hot carrier solar energy converters", *J. Appl. Phys.* **53**, 3813 (1982).

J.H. Werner, S. Kodolinski and H.J.Queisser, "Novel optimization principles and efficiency limits for semiconductor solar-cells", *Phys. Rev. Lett.* **72**, 3851 (1994).

J.H. Werner, R. Brendel and H.-J. Queisser, "Radiative efficiency limit of terrestrial solar-cells with internal carrier multiplication", *Appl. Phys. Letts.* **67**, 1028 (1995).

P. Wuerfel, "Solar energy conversion with hot electrons from impact ionisation", *Solar Energy Mater. Solar Cells* **46**, 43–52 (1997).

Exercises

(The relevant chapters for each question are given in brackets.)

Numerical answers should be given to two significant figures.

The following data may be used:

Boltzmann's constant	$k_B = 1.38 \times 10^{-23}$ J K^{-1}
Electronic charge	$q = 1.60 \times 10^{-19}$ Coulomb
Temperature:	unless otherwise stated, assume
	$T = 300$ K
	$k_B T / q = 0.0258$ eV

Q1. Ideal diode and ideality factor

Two solar cells have an open circuit voltage of 0.85 V and the same short circuit current under a given spectrum. One has an ideality factor of 1, the other has an ideality factor of 2. Use the diode equation to work out the dark saturation current in each case as a fraction of J_{sc}. (You may neglect parasitic resistances). By finding the maximum of the power-voltage characteristic in each case compare the fill factors. How much more efficient is the 'ideal' diode as a photoconverter?

Q2. Cells in parallel and series

A p–n junction solar cell has $V_{oc} = 0.5$ V and $J_{sc} = 20$ mA cm^{-2}. A second cell, of the same area, has $V_{oc} = 0.6$ V and $J_{sc} = 16$ mA cm^{-2}. Assuming that both cells obey the ideal diode equation, find the values of V_{oc} and J_{sc} when the two are connected (a) in parallel and (b) in series.

Q3. Parasitic resistances

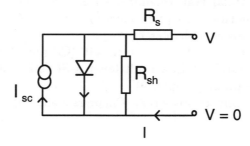

Show, by considering the equivalent circuit above, that the current voltage characteristic for an ideal diode in the presence of parasitic resistances is given by

$$I = I_{sc} - I_0(e^{q(V+IR_s)/kT} - 1) - \frac{V + IR_s}{R_{sh}}$$

where I_0 is the saturation current of the diode, which is assumed ideal, and R_s and R_{sh} represent the series resistance and parallel resistances shown.

(a) By considering the diode equation in the case of high R_s, high R_{sh}, show that $1/R_s$ is approximately equal to the slope of the I–V curve, dI/dV, near $I = 0$. (b) By considering the case of low R_s, low R_{sh}, show that $1/R_{sh}$ is approximately equal to dI/dV near $V = 0$. Explain how the parasitic resistances affect I_{sc} and V_{oc} in each case.

Q4. Ideal diode and light intensity

A solar cell has a short circuit current density of 30 mA cm^{-2} and open circuit voltage of 0.60 V under one-sun illumination at room temperature. Use the ideal diode equation to calculate the open circuit voltage which is expected under illumination by 100 suns, stating any assumptions made. In practice an open-circuit voltage of 0.66 V is measured. Compare this with your result and suggest reasons for any discrepancy.

Q5. Doping and recombination

A semiconductor has an intrinsic carrier density n_i of 2.0×10^{12} m^{-3} at 300 K. A thin slab of this material is exposed to a light pulse of photon energy 2.1 eV and intensity 1000 W m^{-2} for an interval of 1 ns. If the absorption coefficient is 5.0×10^5 m^{-1} at 2.1 eV, calculate the concentration of photogenerated electrons and holes immediately after the pulse. (You may neglect recombination during the light pulse.) Hence find the new total concentration of electrons (n), holes (p) and the product np in the following cases:

(a) if the semiconductor is fully intrinsic (*i.e.* $n = p$ in the dark)?
(b) if it is doped with a concentration of 1.0×10^{22} m^{-3} donor impurities, which may be considered to be fully ionised?

In which case is the rate of radiative recombination faster?

Q6. Quantum efficiency

An $n^{\pm}p$ homo-junction solar cell has emitter thickness of x_n and a base thickness of x_p. If the front surface reflectivity is R and the bulk absorption coefficient is α for photons of energy E,

(a) Find an expression for the photon flux density reaching the base when the incident flux density is $b_o(E)$.
(b) Find an expression for the flux density absorbed in the base.

(c) If each absorbed photon delivers exactly one electron to the contacts, what is the photocurrent density from the cell? You may ignore the emitter photocurrent, and assume that the space charge width is negligible.

(d) Now find an expression for the quantum efficiency (QE).

(e) A $n^{\pm}p$ solar cell has $x_n = 0.5$ μm, $x_p = 20$ μm, and $R = 0.1$ for all photon energies. At 800 nm the absorption coefficient of the cell material is 1.0×10^5 m^{-1}. A student calculates that the QE at this wavelength should be 74%. In fact the measured QE is only 60%. What might be the reason for this?

(f) Use the information above to make a rough estimate of the electron diffusion length in this cell, stating any assumptions which you make.

Q7. Shockley Read Hall recombination

The Shockley–Read–Hall recombination rate is given by

$$U_{\text{SRH}} = \frac{np - n_i^2}{\tau_n(p + p_t) + \tau_p(n + n_t)} \tag{1}$$

where τ_n, τ_p are the electron and hole capture times, and n_t, p_t are given by

$$n_t = n_i e^{(E_t - E_i)/k_B T}$$

$$p_t = n_i e^{(E_i - E_t)/k_B T}$$

where E_t is the energy of the trap level and E_i the intrinsic energy of the semiconductor.

(a) By writing n and p in terms of n_i, E_i and the electron and hole quasi-Fermi levels, E_{F_n} and E_{F_p}, show that U_{SRH} can be written as

$$U_{\text{SRH}} = \frac{n_i \sinh(qV/2k_B T)/\sqrt{\tau_n \tau_p}}{\cosh(\frac{E_{F_n} + E_{F_p} - 2E_i}{2k_B T} + \frac{1}{2}\ln(\frac{\tau_p}{\tau_n})) + e^{-qV/2k_B T}\cosh(\frac{E_t - E_i}{k_B T} + \frac{1}{2}\ln(\frac{\tau_p}{\tau_n}))},$$

(b) Now consider a junction of width w and built-in bias V_{bi} under applied forward bias V ($V < V_{bi}$). Assume that the junction is fully depleted so that E_{F_n} and E_{F_p} are constant, and that E_i varies *linearly* across the junction. Show that the recombination current density due to SRH recombination is given by

$$J_{SRH} = \frac{qn_iw}{\sqrt{\tau_n\tau_p}} \frac{2\sinh(qV/2k_BT)}{q(V_{bi}-V)/k_BT}\xi$$

where

$$\xi = \int_{z1}^{z2} \frac{dz}{z^2 + 2bz + 1}, \quad \text{with } b = e^{-qV/2k_BT}\cosh\left(\frac{E_t - E_i}{k_BT} + \frac{1}{2}\ln\left(\frac{\tau_p}{\tau_n}\right)\right)$$

and the limits of integration are

$$z1 = \sqrt{\tau_p/\tau_n}e^{-q(V_{bi}-V)/2k_BT}, \qquad z2 = \sqrt{\tau_p/\tau_n}e^{q(V_{bi}-V)/2k_BT}.$$

(c) Given that $\int_0^\infty \frac{dz}{z^2+1} = \frac{\pi}{2}$, show that at applied voltages of several times k_BT, ξ can usually be replaced by $\pi/2$ in the expression for J_{SRH} above. Under what conditions would this approximation be invalid?

Q8. Heterojunction

A p–n heterojunction is to be made from two semiconductors, **A** and **B**. Material **A** has band gap E_A, intrinsic carrier density n_{iA} and intrinsic work function ϕ_A (measured from mid-gap to the vacuum level). **B** is an alloy of two materials in the ratio $x{:}1 - x$ and has band gap and intrinsic carrier density given by

$$E_B = E_A + \varepsilon x$$

$$n_{iB} = n_{iA}e^{\varepsilon x/2k_BT}$$

and has intrinsic work function ϕ_B ($\phi_B > \phi_A$) for all compositions x. **B** is to be used for the n-type window layer on a p-type base layer made from **A**.

(a) Find an expression for the value of x at which a step opposing electron collection just begins to appear in the conduction band.

(b) If the doping densities in layers **A** and **B** are N_a and N_d respectively, show that there will always be a net potential gradient driving electrons from **A** to **B** provided that

$$N_aN_d > n_{iA}^2.$$

(c) In a particular case, $E_A = 1.4$ eV, $n_{iA} = 2 \times 10^{12}$ m^{-3}, $\phi_A = 5$ eV, $\phi_B = 5.3$ eV and $\varepsilon = 2.4$ eV. The n-type window layer is designed to admit light of wavelength longer than 500 nm to the p-type base.

   (i) Calculate the size of the step in the conduction band.
   (ii) Sketch the band profile of this heterojunction.
   (iii) Discuss how varying the doping level in the window layer will affect charge collection effeiciency.

Q9. Window layers

A CuInSe$_2$ n–p junction solar cell has a 0.1 μm heavily doped n-type emitter layer and a 4 μm lightly doped p-type base. Use the information below to estimate the relative increase in quantum efficiency at 600 nm when the CuInSe$_2$ emitter layer is replaced with CdS. Explain your reasoning and state clearly any assumptions which you make.

Absorption coefficient of CuInSe$_2$ at 600 nm (α)	1.5×10^7 m^{-1}
Band gap of CdS	2.4 eV
Diffusion length of holes in n^+-type CuInSe$_2$ (L_p)	0.01 μm
Diffusion length of electrons in p-type CuInSe$_2$ (L_n)	2.0 μm

Q10. Electric field in an a-Si p–i–n cell

(a) An amorphous silicon p–i–n solar cell has a p type background doping of 2.0×10^{21} m^{-3} in the intrinsic region and dielectric constant of $\varepsilon = 1.0 \times 10^{-10}$ F m^{-1}. If the cell has a built in bias V_{bi} of 0.9 V, calculate the thickness of the depletion layer in the intrinsic material at zero applied bias, treating the intrinsic region as the p side of a p–n junction. State any approximations which you make.

(b) Would your answer to (a) be a suitable value for the width of the i-region of the solar cell? Give a reason for your answer. Estimate a better value for the i region thickness, stating any assumptions made.

Q11. Current matching in a multilayer a-Si cell

You are asked to design a tandem a-Si solar cell consisting of two p–i–n junctions in series, joined by a tunnel junction. The cell has a total thickness of 1 μm and should be optimized so that the currents are matched for a 600nm light source. If the absorption of a-Si is 3.00×10^6 m^{-1} at 600 nm, how thick should each of the two i regions be? You may assume that the

p and n regions are of negligible thickness, the tunnel junction is ideal and charge collection efficiency is 100% in the intrinsic regions.

Explain what you expect to happen to the short circuit current of the cell if the wavelength of the source is (a) increased and (b) decreased, while keeping the photon flux density the same.

Q12. Light trapping

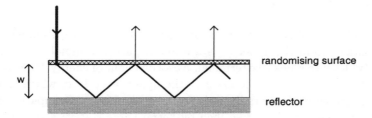

A layer of semiconductor of refractive index n_s and thickness w is mounted between a reflective rear surface and a scattering front surface. The front surface is textured so that it scatters all admitted or reflected light isotropically, whilst still allowing internal light rays approaching the surface at angles smaller than the critical angle to escape. You may take it as given that light rays scattered by the front surface travel a distance of $2w$, on average, before striking the rear surface.

(a) If the rear surface is a perfect reflector, show by considering the path length of rays surviving multiple internal reflections, or otherwise, that average path length of light rays entering the slab is given by

$$\langle l \rangle = 4n_s^2 w \, .$$

(b) If the rear surface is an imperfect reflector with reflectivity R, show that the average path length becomes

$$\langle l \rangle = \frac{2(1 + R)w}{1 - R(1 - 1/n_s^2)}$$

If $n_s = 3$, how large must R be to obtain a path length enhancement of 20?

Q13. Two level photoconverter

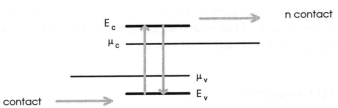

Two level system. Grey arrows indicate processes of electron excitation, relaxation, and transfer to or from contacts.

A molecular photoconverter is modelled as a two level system with loss-less contacts from the upper (c) state to a negative terminal and from the lower (v) level to a positive terminal. Absorbed light of energy $E_g = E_c - E_v$ promotes electrons from the lower to the upper level, from which they are either collected or decay radiatively to the ground state. Occupation of the upper and lower levels is described by Fermi Dirac statistics with occupation functions f_c, f_v and quasi Fermi levels μ_c and μ_v respectively. The output voltage is given by $qV = \mu_c - \mu_v$. The system is intrinsic so that $\mu_c = \mu_v = \frac{1}{2}(E_c + E_v)$ at equilibrium.

For a density of N such molecules in an thin slab of thickness d, the photogenerated current density can be written as

$$J = qN\sigma_0 d \left(G + G_0 - \frac{A}{e^{(E_c - E_v - \mu_c + \mu_v)/k_B T} - 1} \right) (f_v - f_c) \qquad (2)$$

where G_0 is the flux density of photons of energy E_g from the ambient, G the flux density of photons of energy E_g from an illuminating source, σ_0 is the absorption cross section of a single molecule at $T = 0$, f_c, f_v are the occupation probabilities of the upper and lower levels and A is a constant.

(a) Show, by considering the system at equilibrium, that $G_0 = \frac{A}{e^{E_g/k_B T} - 1}$.
(b) Find an expression for the open circuit voltage V_{oc}.
(c) Show, by considering symmetry of the carrier populations, that J can be written as

$$J = qN\sigma_0 d((G + G_0)(x^2 - 1) - A)\frac{1}{(x + 1)^2} \quad \text{where } x = e^{(E_c - \mu_c)/2k_B T}.$$

An additional 'acceptor' level of energy E_c is inserted between the upper level and the n contact, as shown, and an additional 'donor' level at E_v between the lower level and the p contact . The rate of

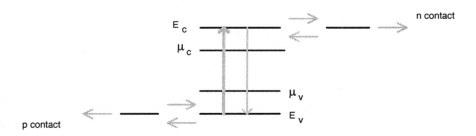

forward electron transfer from the upper level to the acceptor is given by $Kf_c(1 - f_a)$ where f_a is the occupation function of the acceptor level. Electron transfer between donor and lower levels is governed by the same coefficient K. The output voltage is now determined by the quasi Fermi levels of the acceptor and donor levels.

(d) Show that $J = NdK_c(f_c - f_a)$.
(e) Show that V_{oc} is unchanged by the addition of the intermediate levels.
(f) Explain what you expect to happen to J_{sc} as K is varied.

Solutions to the Exercises

Q1. Solution

Ideal diode equation: $J = J_{sc} - J_{0m}(e^{qV/mk_BT} - 1)$

Let K represent the short circuit current density of each cell.

Cell 1. $J_{sc} = K, V_{oc} = 0.85$ V, $\dfrac{J_{01}}{K} = \dfrac{1}{e^{0.85q/k_BT} - 1} = 4.92 \times 10^{-15}$

Cell 2. $J_{sc} = K, V_{oc} = 0.85$ V, $\dfrac{J_{02}}{K} = \dfrac{1}{e^{0.85q/2k_BT} - 1} = 7.01 \times 10^{-8}$

In each case, calculate normalised current,

$$\frac{J}{K} = 1 - \frac{J_{0m}}{K}(e^{qV/mk_BT} - 1)$$

and normalised power

$$\frac{P}{K} = \frac{JV}{K}$$

for V in the range $0 < V < 0.85$ V.

The fill factor is given by

$$\frac{P}{KV_{oc}}.$$

Find that for cell 1, P/K has a maximum of 0.74 V at $V = 0.76$ V and the fill factor is $= 0.87$.

For cell 2, P/K has a maximum of 0.66 V at $V = 0.71$ V and the fill factor is $= 0.78$.

The ideal diode is therefore about 12% more efficient.

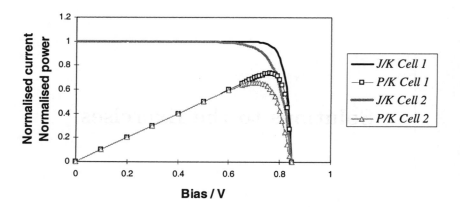

Q2. Solution

We will need the dark saturation current densities. Rearranging the ideal diode equation,

$$J_0 = \frac{J_{sc}}{e^{qV_{oc}/k_BT} - 1}.$$

For cell 1, $J_{01} = \frac{20}{e^{0.5q/k_BT}-1} = 7.66 \times 10^{-8}$ mA cm^{-2}

For cell 2, $J_{02} = \frac{16}{e^{0.6q/k_BT}-1} = 1.27 \times 10^{-9}$ mA cm^{-2}

(a) Cells in parallel. Voltages are equal, currents add. System has

$$V = V_1 = V_2 \qquad J = J_1 + J_2$$

At short circuit, $V = V_1 = V_2 = 0$. Therefore each cell delivers its own short circuit current and

$$J_{sc} = J_{sc1} + J_{sc2} = 20 + 16 = 36 \text{ mA cm}^{-2}$$

At open circuit, $J_1 + J_2 = 0$ and $V_1 = V_2 = V_{oc}$

Therefore

$$J_{sc1} - J_{01}(e^{qV_{oc}/k_BT} - 1) = J_{sc2} - J_{02}(e^{qV_{oc}/K_BT} - 1).$$

Rearrange for V_{oc}:

$$V_{oc} = \frac{kT}{q} \ln \left(\frac{J_{sc1} + J_{sc2}}{J_{01} + J_{02}} + 1 \right)$$

Substituting numerical values, get

$$V_{oc} = 0.52 \text{ V}$$

(b) Cells in series. Currents are equal, voltages add. System has

$$V = V_1 + V_2 \qquad J = J_1 = J_2$$

At open circuit, $J = J_1 = J_2 = 0$. Therefore each cell generates its independent V_{oc} and

$$V_{oc} = V_{oc1} + V_{oc2} = 0.5 + 0.6 = 1.1 \text{ V}$$

At short circuit, $V_1 + V_2 = 0$ and $J_1 = J_2$

Let $x = e^{qV_1/k_B T}$ and $x^{-1} = e^{qV_2/k_B T}$
Then

$$J_{sc1} - J_{01}(x - 1) = J_{sc2} - J_{02}\left(\frac{1}{x} - 1\right).$$

Solving the quadratic for x,

$$x = \frac{(J_{sc1} - J_{sc2} + J_{01} - J_{02}) \pm \sqrt{(J_{sc1} - J_{sc2} + J_{01} - J_{02})^2 + 4J_{01}J_{02}}}{2J_{01}}$$

whence $x \approx \frac{(J_{sc1} - J_{sc2})}{J_{01}}$ where we have used the fact that J_{01}, $J_{02} \ll J_{sc1}$, J_{sc2}.
This yields

$$J_{sc} = J_{sc1} - J_{01}(x - 1) \approx J_{sc2}$$

i.e. the short circuit current density approximates to that of the individual cell with the lower J_{sc}.

Q3. Solution

(i) High R_s, high R_{sh}
    We can neglect the term involving R_{sh}.
    Then

$$I(V) = I_{sc} - I_o(e^{q(V+IR_s)/k_B T} - 1).$$

Rearranging to make V the subject

$$V = \frac{k_B T}{q} \ln\left(\frac{I_{sc} + I_o - I}{I_o}\right) - IR_s.$$

Differentiating

$$\frac{dV}{dI} = -\frac{k_B T}{q}\left(\frac{1}{I_{sc} + I_o - I}\right) - R_s.$$

At the open-circuit point $(I = 0, V = V_{oc})$

$$R_s = -\frac{dV}{dI}\bigg|_{V=V_{oc}} - \frac{k_B T}{q}\left(\frac{1}{I_{sc} + I_o}\right).$$

The second term is small — *e.g.* compared to the characteristic resistance of the cell since $(k_B T/q)$ is much smaller than V_{oc} (and also then the operating voltage V_m). So, whenever the series resistance becomes large compared with $(k_B T/q I_{sc})$ it is approximately equal to (dV/dI) evaluated at $V = V_{oc}$.

Cells with a high series resistance have a noticeable gradient in $I(V)$ near the open circuit point. R_s has no effect on V_{oc} since the term IR_s in the diode equation vanishes at open circuit. High R_s may reduce J_{sc}.

(ii) **Low R_s, low R_{sh}**
We can neglect R_s.
The diode equation becomes

$$I(V) = I_{sc} - I_o(e^{qV/k_B T} - 1) - \frac{V}{R_{sh}}.$$

No need to rearrange since this is explicit for I.

Differentiating,

$$\frac{dI}{dV} = -\frac{q}{k_B T}I_o e^{qV/k_B T} - \frac{1}{R_{sh}}.$$

At the short-circuit point $(V = 0, I = I_{sc})$

$$\frac{1}{R_{sh}} = -\frac{dI}{dV} + \frac{q I_o}{k_B T}.$$

In the absence of leakage currents $(R_{sh} \to \infty)dI/dV$ is equal to the second term on the right-hand side of the above expression and is very small. For substantial leakage currents, the second term is negligible compared to $dI/dV R_{sh} \approx (dV/dI)$ evaluated at $V = 0$. Note that R_{sh} influences the $I(V)$ curve most strongly at $V = 0$. Cells with a low R_{sh} have a noticeable negative gradient in $I(V)$ near $V = 0$.

R_{sh} has no effect on I_{sc} since the term V/R_{sh} in the diode equation vanishes at short circuit. Low R_{sh} may reduce V_{oc}.

Q4. Solution

Ideal diode equation:

$$J = J_{sc} - J_0(e^{qV/k_BT} - 1)$$

At open circuit, $J = 0$ and

$$V_{oc} = \frac{k_BT}{q} \ln\left(\frac{J_{sc}}{J_0} + 1\right).$$

Under one sun,

$$V_{oc}(1) = \frac{k_BT}{q} \ln\left(\frac{30}{J_0} + 1\right)$$

where J_0 is in mA cm^{-2}

Under 100 suns,

- Assume that J_{sc} increases linearly with concentration. Therefore J_{sc} becomes 3000 mA cm^{-2}
- Assume that J_0 does not change with concentration.
- Assume that fill factor does not change.

Then expect

$$V_{oc}(100) = \frac{k_BT}{q} \ln\left(\frac{3000}{J_0} + 1\right) \approx V_{oc}(1) + \frac{k_BT}{q} \ln(100).$$

substituting numerical values,

$$V_{oc}(100) = 0.60 + 0.12 = 0.72 \text{ V}$$

The lower actual V_{oc} of 0.66 V under 100 suns could be due to:

(i) heating of the cell under concentration, causing J_0 to increase.
(ii) sublinear increase in J_{sc} with concentration due to series resistance. (This will reduce the fill factor.)

Q5. Solution

Photon flux density:

$$b = \frac{1.0}{2.1 \times q} = 2.97 \times 10^{21} \ \mathrm{m^{-2}s^{-1}}$$

In a thin slab, assume uniform photogeneration rate of αb.

Photogenerated carrier density in time t (neglecting recombination):

$$n_{\mathrm{g}} = \alpha b t = 5.0 \times 10^{-5} \times 2.97 \times 10^{21} \times 1 \times 10^{-9} \ \mathrm{m^{-3}} = 1.49 \times 10^{18} \ \mathrm{m^{-3}}$$

Immediately after laser pulse, electron density $n = n_0 + n_{\mathrm{g}}$ and hole density $p = p_0 + n_{\mathrm{g}}$

Case (a): Intrinsic semiconductor.

$$n_0 = p_0 = n_{\mathrm{i}} = 2 \times 10^{12} \ \mathrm{m^{-3}}$$

$$n = p = 1.49 \times 10^{18} + 2.0 \times 10^{12} \approx 1.49 \times 10^{18} \ \mathrm{m^{-3}}$$

$$np = 2.21 \times 10^{36} \ \mathrm{m^{-6}}$$

Case (b): Doped semiconductor with $N_{\mathrm{d}} = 10^{22} \ \mathrm{m^{-3}}$.

$$n_0 = 1.0 \times 10^{22} \ \mathrm{m^{-3}} \qquad p_0 = \frac{n_{\mathrm{i}}^2}{n_0} = 4.0 \times 10^2 \ \mathrm{m^{-3}}$$

$$n = 1.49 \times 10^{18} + 1.0 \times 10^{22} \approx 1.0 \times 10^{22} \ \mathrm{m^{-3}}$$

$$p = 1.49 \times 10^{18} + 4.0 \times 10^2 \approx 1.49 \times 10^{18} \ \mathrm{m^{-3}}$$

$$np = 1.49 \times 10^{40} \ \mathrm{m^{-6}}$$

Radiative recombination is proportional to the product np. Therefore radiative recombination will be faster for the doped semiconductor.

Q6. Solution

(a) Photon flux density reaching the base is: $b_0(E)(1 - R)e^{-\alpha x_{\mathrm{n}}}$

(b) Photon flux density absorbed in the base is:

$$b_0(E)(1 - R)e^{-\alpha x_{\mathrm{n}}}(1 - e^{-\alpha x_{\mathrm{p}}})$$

(c) Assuming that each photon absorbed in the base results in one electron delivered to the external circuit, then the photocurrent density due to photons of energy E is given by

$$J = qb_0(E)(1 - R)e^{-\alpha x_n}(1 - e^{-\alpha x_p})$$

(d) Quantum efficiency = current density collected $/(q\times$ incident photon flux density)

$$QE(E) = \frac{qb_0(E)(1 - R)e^{-\alpha x_n}(1 - e^{-\alpha x_p})}{qb_0(E)} = (1 - R)e^{-\alpha x_n}(1 - e^{-\alpha x_p})$$

(e) Substituting numerical values, the result is

$$QE = 0.9 \times e^{-0.05} \times (1 - e^{-2}) = 0.74.$$

The measured QE is 0.6. The reason for lower quantum efficiency could be incomplete collection of carriers in the base due to short electron diffusion length, rear surface recombination or junction recombination.

(f) Suppose the discrepancy in (e) is due to short electron diffusion length in the base. Then, if we assume that only carriers photogenerated within an electron diffusion length, L_n, of the junction are collected, the QE can be written

$$QE(E) = (1 - R)e^{-\alpha x_n}(1 - e^{-\alpha L_n}).$$

Setting $QE = 0.6$ and solving for L_n,

$$e^{-\alpha L_n} = 1 - \frac{0.6}{0.9 \times 0.951} = 0.299$$

$$L_n = 12 \ \mu m$$

Q7. Solution

(a) Write $n = n_i e^{(E_{Fn} - E_i)/k_B T}$, $p = n_i e^{(E_i - E_{Fp})/k_B T}$ and substitute into Eq. (1). Use the fact that $e^{(E_{Fn} - E_{Fp})/k_B T} = e^{qV/k_B T}$ and multiply top and bottom by $e^{-qV/2k_B T}$ Multiply top and bottom of right-hand side by $\sqrt{\tau_n/\tau_p}$ and write

$$\sqrt{\frac{\tau_n}{\tau_p}} = \exp\left(\frac{1}{2}\ln\left(\frac{\tau_n}{\tau_p}\right)\right).$$

Write numerator as sinh() and collect terms in denominator into cosh() expressions. This yields the given result.

(b) At applied bias V, the potential dropped across the junction is $V_{bi} - V$. E_i should increase linearly by an amount $q(V_{bi} - V)$ from the n side ($x = -w/2$) to the p side ($x = w/2$). Choose zero of E_i midway between quasi Fermi levels, so

$$E_i = \frac{1}{2}(E_{F_n} + E_{F_p}) + \frac{q(V_{bi} - V)x}{w}.$$

Then, finding the recombination current by integrating the continuity equation

$$J_{SRH} = q \int_{-w/2}^{w/2} U_{SRH} dx$$

$$= \frac{qn_i \sinh(qV/2k_BT)}{\sqrt{\tau_n \tau_p}} \int_{-w/2}^{w/2} \frac{dx}{\cosh(\frac{q(V_{bi}-V)x}{wk_BT} + C) + b}$$

where $C = \frac{1}{2}\ln(\frac{\tau_n}{\tau_p})$, and b is given in the question.

Make the substitution,

$$z = \exp\left(\frac{q(V_{bi} - V)x}{wk_BT} + C\right)$$

whence

$$J_{SRH} = \frac{2qn_i w \sinh(qV/2k_BT)}{\sqrt{\tau_n \tau_p} q(V_{bi} - V)/k_BT} \times \int_{z1}^{z2} \frac{dz}{z^2 + 2bz + 1}$$

with $z1$, $z2$ as given in the question.

(c) We can replace ξ with $\pi/2$ in the limit where $b \ll 1$. From the above, assume that

$$b = e^{-qV/2k_BT} \cosh\left(\frac{E_t - E_i}{k_BT} + \frac{1}{2}\ln\left(\frac{\tau_p}{\tau_n}\right)\right).$$

At biases where $qV \gg 2k_BT$, the factor $e^{-qV/2k_BT}$ becomes $\ll 1$, so that $b \ll 1$ overall provided that the second factor is not large.

However, b can still be significant at operating biases if

$$\cosh\left(\frac{E_t - E_i}{k_BT} + \frac{1}{2}\ln\left(\frac{\tau_p}{\tau_n}\right)\right) \approx e^{qV/2k_BT}.$$

This happens in the limits where the trap energy is far from mid gap: $|E_t - E_i| \gg k_BT$ and where the carrier lifetimes are very different $\tau_n \gg \tau_p$ or $\tau_p \gg \tau_n$.

Q8. Solution

(a) The step in the conduction band is due to differences in the electron affinity.

Electron affinity of **A** $\quad \chi_A = q\phi_A - \dfrac{1}{2}E_A$

Electron affinity of **B** $\quad \chi_B = q\phi_B - \dfrac{1}{2}(E_A + \varepsilon x)$

The step opposes electron drift from **A** to **B** when $\chi_B < \chi_A$. This happens when

$$q\phi_B - \frac{1}{2}(E_A + \varepsilon x) < q\phi_B - \frac{1}{2}E_A$$

i.e. when,

$$x > \frac{2q(\phi_B - \phi_A)}{\varepsilon}.$$

(b) The net potential drop across the junction is determined by the difference in conduction band levels far from the junction. For the junction at equilibrium, the conduction band levels can be measured from the Fermi level, E_F:

Conduction band level in **A**, far from junction:

$$E_{cbA} - E_F = \frac{1}{2}E_A + k_B T \ln\left(\frac{N_a}{n_{1A}}\right)$$

Conduction band level in **B**, far from junction:

$$E_{cbB} - E_F = \frac{1}{2}E_B - k_B T \ln\left(\frac{N_d}{n_{iB}}\right)$$

Substituting for E_B, n_{iB},

$$E_{cbB} - E_F = \frac{1}{2}(E_A + \varepsilon x) - k_B T \ln\left(\frac{N_d}{n_{iA}}\right) + k_B T \ln(e^{-\varepsilon x/2k_B T})$$

$$= \frac{1}{2}E_A - k_B T \ln\left(\frac{N_d}{n_{iA}}\right)$$

For a net potential gradient driving electrons from **A** to **B**, we want $E_{cbA} > E_{cbB}$.

Using expressions for E_{cbA}, E_{cbB}, this implies

$$k_B T \ln \left(\frac{N_a N_d}{n_{iA}^2} \right) > 0$$

whence

$$N_a N_d > n_{iA}^2$$

as required.

(c) If **B** just admits light of 500 nm, the band gap of **B** is $E_B = 1240/500 = 2.48$ eV

(i) Step in conduction band:

$$\chi_A - \chi_B = q(\phi_A - \phi_B) - \frac{1}{2}(E_A - E_B)$$
$$= 5.0 - 5.3 - 5.0(1.4 - 2.48) = 0.24 \text{ eV}$$

This is significant: a step height many times greater than $k_B T$ means thermal emission over the step will be slow.

(ii)

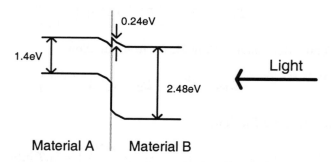

(iii) Increasing N_d will increase the band bending on the n side of the junction, and so reduce the thickness of the barrier opposing electron flow and increase collection via tunnelling. Reducing N_d will make the barrier thicker and reduce tunnelling. For highest carrier collection, **B** should be highly doped.

Q9. Solution

Assume that carriers generated within a diffusion length of the junction contribute to the photocurrent. Let b_0 represent the incident flux.

For the CuInSe$_2$ homojunction cell,

- flux lost in emitter is: $b_0 e^{-\alpha(x_n - L_p)}$
- flux absorbed within a diffusion length of junction:

$$b_0 e^{-\alpha(x_n - L_p)} \left(1 - e^{-\alpha(L_n + L_p)}\right)$$

- photocurrent density: $J = q b_0 e^{-\alpha(x_n - L_p)} \left(1 - e^{-\alpha(L_n + L_p)}\right)$
- quantum efficiency: $QE = e^{-\alpha(x_n - L_p)} \left(1 - e^{-\alpha(L_n + L_p)}\right)$

For the CdS/CuInSe$_2$ heterojunction cell,

- longest wavelength absorbed by CdS $= 1240/2.4 = 516$ nm.
- CdS should be transparent to 600 nm light, threfore assume that no light is absorbed in the emitter.

Then

- flux absorbed within a diffusion length of junction: $b_0(1 - e^{-\alpha L_n})$
- quantum efficiency: $1 - e^{-\alpha L_n}$

$$\frac{\text{heterojunction } QE}{\text{homojunction } QE} = \frac{1 - e^{-\alpha L_n}}{e^{-\alpha(x_n - L_p)}(1 - e^{-\alpha(L_p + L_n)})}$$

- Substituting numerical values: $\alpha = 1.5 \times 10^7$ m^{-1}, $L_n = 2 \times 10^{-6}$ m, $L_p = 1 \times 10^{-8}$ m, find

$$\text{heterojunction } QE = \text{homojunction } QE \times 3.86.$$

Therefore replacing the CuInSe$_2$ window by CdS increases the QE at 600 nm by a factor of 3.9.

Q10. Solution

(a) The depletion width of a p–n junction at zero bias is given by

$$W = \sqrt{\frac{2\varepsilon_s V_{bi}}{q} \left(\frac{1}{N_a} + \frac{1}{N_d}\right)}.$$

In this case, replace N_a with N_i for the background doping of the p-type intrinsic region, and assume that the n region is doped such that $N_d \gg N_i$.

Then the entire width is dropped within the intrinsic region and

$$W = \sqrt{\frac{2\varepsilon_s V_{bi}}{qN_i}}.$$

Substituting numerical values we get $W = 0.75$ μm.

(b) Part (a) shows that the electric field will fall to zero at the p–i interface of an intrinsic region 0.75 μm thick at zero applied bias. Under operating conditions, the bias dropped across the junction is reduced to $(V_{bi} - V)$. This reduces the depleted width, so that an intrinsic region 0.75 μm thick will be partly undepleted at any $V > 0$, and will not function effectively as a solar cell where collection from the i-region requires electric field. To estimate a better i-region thickness, replace V_{bi} with $(V_{bi} - V)$ in the expression for W. Estimating an operating bias of 0.6 V, we get $W = 0.43$ μm.

Q11. Solution

Since the cells are connected in series, it is necessary to match the currents from the front and back cell. Now since less light reaches the back cell, the back cell must be thicker to produce the same photocurrent. The optimum ratio of thicknesses will depend on illumination conditions.

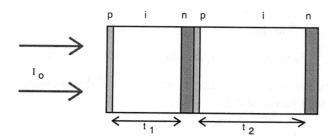

Let the first i layer be of thickness t_1 and the second of thickness t_2. Suppose the photon flux density entering the first i layer is b_0. Then the total absorbed flux in that layer is $b_0 \int_0^{t_1} e^{-\alpha t} dt$. This is proportional to the photocurrent.

The photon flux density entering the second i layer is $b_0 e^{-\alpha t_1}$. Then the total absorbed flux there is $I_0 e^{-\alpha t_1} \int_0^{t_2} e^{-\alpha t} dt$. This is proportional to the photocurrent from that layer.

If the photocurrents are equal then,

$$I_0 \int_0^{t_1} e^{-\alpha t} dt = I_0 e^{-\alpha t_1} \int_0^{t_2} e^{-\alpha t} dt$$

$$-\frac{I_0}{\alpha} [e^{-\alpha t_1} - 1] = -\frac{I_0}{\alpha} e^{-\alpha t_1} [e^{-\alpha t_2} - 1]$$

$$1 - e^{\alpha t_1} = e^{-\alpha t_2} - 1$$

$$e^{\alpha t_1} + e^{-\alpha t_2} = 2$$

Substituting numerical values and using $t_1 + t_2 = 1$ μm, get

$$e^{3t_1} + e^{-3(1-t_1)} = 2$$

$$e^{3t_1} = \frac{2}{1 + e^{-3}} = 1.905$$

whence

$$t_1 = 0.215 \ \mu\text{m}$$

$$t_2 = 0.785 \ \mu\text{m}$$

(a) Absorption of a seminconductor generally increases with decreasing wavelength in the visible. Therefore, if the wavelength of the source light is increased, the absorption coefficient should decrease, resulting in smaller absorbed flux in the first layer and larger in the second layer. The currents from the two layers must match, so the short circuit photocurrent will be limited by the short circuit photocurrent from the first layer. In order to match the current from the second layer with that from the first, the tunnel junction will become positively charged to forward bias the second junction.

(b) If the wavelength of the source light is decreased, the absorption coefficient should increase, resulting in larger absorbed flux in the first layer and smaller in the second layer. Now the short circuit photocurrent is limited by the short circuit photocurrent from the second layer. The tunnel junction becomes negatively charged to forward bias the first junction and reduce the current from that layer to match the second.

Q12. Solution

(a) Consider the light path in terms of the number of double passes. The path length for each double pass is $2 \times 2w = 4w$. The probability of escaping after only one double pass, *i.e.* with path length $4w$, is $1/n_s^2$. The probability of surviving the first bounce but escaping after the second is $1/n_s^2(1 - 1/n_s^2)$; this path has length $8w$. Summing up the length for each path times the probability of surviving for exactly that number of passes yields the following expression for the average path length:

$$\langle l \rangle = 4w\frac{1}{n_s^2} + 8w\frac{1}{n_s^2}\left(1 - \frac{1}{n_s^2}\right) + 12w\frac{1}{n_s^2}\left(1 - \frac{1}{n_s^2}\right)^2 + \cdots$$

$$= \frac{4w}{n_s^2}\sum_1^\infty i\left(1 - \frac{1}{n_s^2}\right)^{i-1}$$

Using the result that

$$\sum_0^\infty x^i = \frac{1}{1 - x} \qquad \text{for } x < 1,$$

we have

$$\sum_1^\infty ix^{i-1} = \frac{1}{(1 - x)^2} \qquad (x < 1).$$

Substituting $(1 - \frac{1}{n_s^2})$ for x and simplifying we have

$$\langle l \rangle = \frac{4w}{n_s^2}\sum_1^\infty i\left(1 - \frac{1}{n_s^2}\right)^{i-1} = 4n_s^2 w$$

(b)

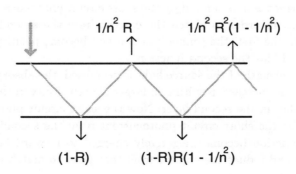

Adding up the contributions from different numbers of passes, now including odd numbers of passes to allow for escape through the rear surface we have

$$\langle l \rangle = 2w(1-R)+4w\frac{1}{n_s^2}R+6w(1-R)R\left(1-\frac{1}{n_s^2}\right)+8w\frac{1}{n_s^2}R^2\left(1-\frac{1}{n_s^2}\right)+\cdots$$

collecting the terms representing odd numbers of passes and even numbers of passes separately we have

$$\langle l \rangle = 2w(1-R)\sum_{1}^{\infty}(2i-1)X^{i-1} + 4w\frac{1}{n_s^2}R\sum_{1}^{\infty}iX^{i-1}$$

where

$$X = R\left(1-\frac{1}{n_s^2}\right).$$

Using the results for the summed series given in the solution to (a), substituting for X and simplifying we have

$$\langle l \rangle = \frac{2w(1+R)}{1-R(1-\frac{1}{n_s^2})} \qquad \text{as required}$$

(c) Rearranging to make R the subject,

$$R = \frac{1-2w/\langle l \rangle}{1-1/n_s^2+2w/\langle l \rangle}$$

A path length enhancement of 20 means that $\frac{w}{\langle l \rangle} = 0.05$
  Substituting numerical values,

$$R = \frac{1-2\times 0.05}{1-1/3^2-2\times 0.05} = 0.91$$

Q13. Solution

(a) At equilibrium, only thermal photons are present ($G = 0$), the quasi Fermi levels are equal and $J = V = 0$.
  Setting $J = 0$, $\mu_c = \mu_v$ and $G = 0$ in (1), we get

$$G_0 = \frac{A}{e^{(E_c-E_v)/k_B T} - 1}.$$

(b) Output voltage is due to difference in μ_c and μ_v. At open circuit,

$$J = 0 \quad \text{and} \quad qV_{oc} = \mu_c - \mu_v.$$

Substituting into (1),

$$G + G_0 - \frac{A}{e^{(E_g - qV_{oc})/k_B T} - 1} = 0$$

Rearranging for V_{oc},

$$V_{oc} = \frac{E_g}{q} - \frac{k_B T}{q} \ln\left(\frac{A}{G + G_0} + 1\right)$$

(c) The system is intrinsic, so $f_c = 1 - f_v$ in equilibrium. Moreover, since carriers can only be generated or destroyed in pairs, and both contacts are ideal, this is true also under photogeneration. Therefore $E_c - \mu_c = -(E_v - \mu_v)$ and we can write

$$x = e^{(E_c - \mu_c)/k_B T} = e^{-(E_v - \mu_v)/k_B T}.$$

Then

$$f_c = \frac{1}{x+1} \qquad f_v = \frac{x}{x+1}$$

and

$$e^{(E_c - E_v - \mu_c + \mu_v)/k_B T} = x^2.$$

Substituting into (1), we obtain the given form for J.

(d) At equilibrium, $\mu_c = \mu_a$ whence $f_c = f_a$

If the rate of electron transfer from upper state to acceptor level is $Kf(1 - f_a)$, then the rate of electron transfer from acceptor to upper state must be $Kf_a(1 - f_c)$, to ensure where is no net electron transfer at equilibrium. Away from equilibrium, the net transfer rate is $Kf_c(1 - f_a) - Kf_a(1 - f_c) = K(f_c - f_a)$.

For a density of N molecules in a slab of thickness d this produces a current density $J = qNdK(f_c - f_a)$ as required.

(e) Terminal voltage is determined by the quasi Fermi levels of the contact levels, *i.e.* by μ_a and μ_d.

At open circuit, $J = 0$ therefore $f_c = f_a$ (from part (d)) and $\mu_c = \mu_a$. Similarly, $\mu_v = \mu_d$ Therefore V_{oc} is determined by μ_c only; can use same expression as in (b).

(f) As $K \to \infty$, (limit of ideal contact), $\mu_c \to \mu_a$ and $\mu_v \to \mu_d$ even under current flow. Then have $\mu_c = \mu_v$ at short circuit, and Jsc tends to its maximum value,

$$J = qN\sigma_0 dG \left(1 - \frac{2}{e^{(E_c - E_v)/2k_B T} - 1}\right)$$

As $K \to 0$, carrier collection from the upper and lower levels becomes slow compared to radiative recombination. The fraction of carriers lost to radiative recombination increases, so that the fraction collected falls and J_{sc} reduces.

Index